God's Universe *in* Four Dimensions

Les Burgess

Order this book online at www.trafford.com
or email orders@trafford.com

Most Trafford titles are also available at major online book retailers.

Printed in the United States of America.

ISBN: 978-1-4269-5863-2 (sc)
ISBN: 978-1-4269-5864-9 (e)

Trafford rev. 03/10/2011

www.trafford.com

North America & international
toll-free: 1 888 232 4444 (USA & Canada)
phone: 250 383 6864 • fax: 812 355 4082

Content

show (or imagine) the universe in one picture because of the change of time-frame with distance.

Chapter 4 **47**

How divinity works A finite universe favors belief No absolute motion in space, only relative motion Everlasting universe – like God Miracles Metaphysical view of reality Theodor Kaluza and extra dimensions

Chapter 5 **54**

God reveals himself to the writer with four short words: "Thou shalt not kill" spoken angrily in the writer's mind. He had been seeking solutions to the world's problems and the simple solution was to destroy all the bad people, but that is not God's plan until the 'Day of Judgment' God clearly identifies himself with Nature to Moses (Exodus 4.11) Nature's order equates to God's order 'Intellectuals' corrupting the Church Man's connection through evolution to animals cause of much sin

Chapter 6 **60**

Hell on earth The End is near Sensibility and brutish ignorance Old Testament for Jewish nation New Testament for individual salvation.

Chapter 7 **68**

About the author Spelling reform Logic of God and nature Kingdom of God. Judgmental attitude Huge potential of the universe Have faith in God Christians should be ladies and gentlemen

Einstein's relativity theory finally explained. Author reveals astounding shape of everlasting finite boundless universe

Chapter 1

The chapters are generally short and each has a brief description of the subject matter. This new theory of the universe 'breakthrough' (because that is what it is) should be invaluable because it sheds light on apparent anomalies in astronomy and other unanswered questions in physics, such as the instantaneous or infinite velocity of light in its own time-frame, why the red shift is misleading (caused by constant curvature of space-time - the universe is not expanding), the increase in mass of an accelerating object, the contraction of time in relativity theory, the 'spooky' velocity of light that may throw further light on mysterious quantum (sub atomic) behavior, and UFO's, believe it or not. It also presents a compatible design of the universe with Einstein's relativity theory that endless 3-dimensional space doesn't.

A comprehensive understanding of the universe begins with Einstein's theory of relativity and its relevant transformations that, incidentally, do not require the reader to have the sophisticated mathematical knowledge that Einstein used. We only need the simple to understand conclusions, which can, and have been, verified. Astronomical observations at the limit of observable space can be misleading because a 4-dimensional universe has a vastly different shape to a 3- dimensional one. We are looking along a curved line of sight bending downward a hundred and eighty degrees in every direction at the limit of visibility, the cosmic horizon. Thus, an easterly line of sight will take us nearly to the limit of a westerly line of sight. Also the force of gravity would follow the same path. This would change the basics of the gravitational collapse of all the matter in the universe fear that caused Einstein so much trouble. The universe won't collapse. It isn't expanding. It never began and it will never end. It is a beautiful logical system with an omniscient, omnipotent, logical and interacting God. The only 'end' will be when 'Judgment Day' arrives unexpectedly, probably in the near future. The writer doesn't know the date, but he knows something of how God works.

It is known that every supernova of a certain size (10 to 20 times the mass of our sun) radiates light with the same intensity. This creates what astronomers call a 'standard candle'. There are others but supernovas are the best because they are the brightest. "By comparing how much dimmer the supernovas actually do appear relatively to distance, two teams of astronomers figured they could determine how much the expansion of the universe was slowing down. But to their surprise, when they looked as far as halfway across the universe, six or seven billion light-years away, they found that the supernovas were dimmer than expected. "The two teams both concluded that the expansion of the universe isn't slowing down. It's speeding up." (35 Oct/Nov

2010 'Cosmos' magazine (Australia) p. 71-72). This has been confirmed by other astronomers. But this conclusion is based on conventional 3-dimensional thinking.

These observations support exactly what the writer (and Einstein's theory, properly understood) is saying about the universe. Chapters 2 and 3 explain the theory more fully, but it may take a day or two to stretch the imagination to see the truth of it – which is why it hasn't been understood (or believed) before now. The writer has spent a long lifetime pondering the question of the origin of the universe – and other things.

Light normally diminishes by the inverse square law. The same light twice as far away will only be a quarter as bright. In a spherical 4-dimensional universe the rate of diminishment would, in addition, be exponential, increasing slowly at first with increasing distance then more and more rapidly to infinity at the visual finite (cosmic) horizon. The cosmic horizon, beyond which we can see nothing, would be near the theoretical diameter of the sphere from the observer's point of view at the center – every observer appears to be at the center, but only from his point of view. This is only one of the surprises in the 4-d universe. Ordinary 3-dimensional geometry does not work in space-time except over 'short' (in astronomical terms) distances.

Beyond about three billion lightyears the 4-dimensional effects become more noticeable. In the last half-billion they increase to infinity. This is because spacetime at this point, about 10 billion lightyears, has curved 180 degrees relative to the viewer – and likewise from the observed point of view. If we could see farther we could see ourselves as we were 20 billion years ago, but we were only formed (or reformed) four and a half billion years ago. We would therefore see whatever occupied the space at that time.

The universe behaves in one sense like a sphere (as Einstein said it should) but progresses along the time continuum (time passes and natural changes take place). A space traveller moves in an apparently (to him) straight line, but which is really very slightly (because of the enormous size of the universe) curved which will eventually bring him back to the space he left but not the time. If he travelled at the speed of light it would still take him 20 billion years in the universal time frame, but no time at all in his own time frame. 4 dimensions cannot be imagined in one 3-dimensional picture because the time factor requires a changing scene as point of view changes. Time changes with place or point of view.

Attempting to view the universe in one-picture results in a highly distorted view of very distant galaxies, though it works very well over short distances. The time contraction to the traveller moving at the speed of light is one of Einstein's transformations and has been proved experimentally with clocks. For the sake of simplification our imaginary spaceship moves immediately at the speed of light but, like everything else that moves or accelerates, only in its own time-frame.

It is this distorted view that we see when we look through a sufficiently powerful telescope at the limits of observation. We cannot 'see' the time part of Einstein's space-time continuum, we can only see its effect (displacement) on galaxies in space at great distances. The time dimension is evidently just as 'real' as the space dimensions. The limit of observation is the cosmic horizon that works like the horizon on earth that moves as the observer moves. A lightyear is the distance light travels in a year at 300,000 km per second.

Note that the visible diameter of the 4-dimensional spherical universe extends around half the circumference because of the curvature of spacetime into a circle, but it stops there. In the process of curving its wavelength becomes attenuated exponentially (to the observer) resulting in the cosmic horizon, which moves as the observer moves. Also note that the light that we see by commenced from the other side of the universe and became invisible by the time it arrived here, so we can't see quite the full distance. What we can see is a few very dim quasars then nothing beyond. There is a small (astronomically speaking) difference between the theoretical cosmic horizon and the practical one.

This is important to astronomers because this is the region that produces constant 'background radiation' of very low temperature that appears to confirm the 'big bang' theory, which never happened. The other important thing to note is the path of the light is curved into a semicircle, not a straight line as astronomers evidently think. Upon this important point depends the survival of the universe, and Einstein's peace of mind, because lines of force (gravity) are curved 180 degrees preventing the gravitation collapse of the spherical universe. Einstein proved his theory mathematically but mathematics didn't reveal what was needed to show that straight-line forces don't exist in 4-dimensions except over relatively short distances. He did say that 'intuition was necessary to progress further'.

The writer (who was born in 1926) has been fascinated by Einstein's theory all his life but only made partial progress in pondering the origin of the universe until very recently. Of course, the universe is, like God, immortal, both in theory and in fact by observation. The reader will be glad to know that the universe is very much 'alive' and full of energy.

What the aforementioned astronomers should have been concentrating on was the rate of diminishment of light relative to distance. Was the high rate of diminishment due to higher speed of recession or was it due to something else, such as the writer asserts, the curvature of space-time into a spherical universe? If the rate of diminishment is found to be increasingly exponential to infinity forming a moveable cosmic horizon at a finite distance then the universe should be seen as a 4-dimensional boundless sphere as Einstein predicted.

'Boundless' in Einstein's space-time means we cannot see the boundary, approach it or cross it because it doesn't exist in those terms. The cosmic horizon 'marks' it to an observer from across the universe, but when it is approached it moves away like the horizon on earth. Although the universe is boundless we cannot leave it. We cannot approach the (theoretical) boundary because every observer sees himself, and is, always at the center of his view of the universe.

The writer has explained in the next chapters how such a universe behaves in the eyes of its viewers. The reason why Einstein did not progress further with his theory was because he thought it would collapse under the force of gravity, which is also why other scientists haven't embraced the whole idea.

The tiny quantum world of the atom seems to obey a different set of rules to the everyday world of space and time. The behavior of electrons (which appear to have orbits around the very small but very dense nucleus of atoms) is one of the great mysteries of nature because of what is called 'Heisenberg's Uncertainty Principle'. This states that the electron is a particle that can be described as a wave, but we can "know the path an electron takes…or we can know where it is at a given instant but we cannot know both" (from p. 188 in Bill Bryson's 'A Short History of Nearly Everything').

The writer suggests that nature has changed the space-time rule in the quantum world and left out the time element. It would be impossible to know the position of an electron if it moved at infinite speed (instantaneously) in the observer's time frame. Note: Radiant energy always travels instantaneously but only in its own time frame, not the observer's. The conclusion is that the atomic electron is behaving normally but the observer's time frame in the quantum spatial dimensions isn't. Atoms seem to have their own space relative only to themselves. Another dimension perhaps? If we look into it we find things happening at infinite speed, but it is not the speed of photons or electrons that has changed, it is the space with the time factor absent: 3-dimensional space. It is 4-dimensional space that limits the speed of light, because it moves over space that has different time frames relative to distance travelled – one second to every 300,000 km. Further discussion of the shape of the universe with descriptions of maps in 2 and 3 dimensions continues in chapter 2.

It might interest the reader to know that atoms are mostly empty space. The nucleus is one millionth of a billionth of the volume of an atom but contains almost all of the atom's mass. It is made up of positively charged protons and uncharged neutrons which are tightly bound together. The light negatively charged electrons (very small mass) orbit the nucleus but only in strictly controlled paths called orbits that (somehow) prevent them from being drawn into the nucleus. It is the like electrical charges that prevent 'solid' objects (like billiard balls) from passing through each other without touching (like electrical charges repel; unlike charges, positive and negative, attract).

If the writer disappoints some scientific people by writing about religion, regarding it as superstition and unconnected with science, once again they will need to extend their minds. The metaphysical question, 'Is there a mind behind the universe?' is valid. After all we have minds and we are only made of common matter. If nature can produce us (according to current scientific belief) the potential might exist to produce God – or vice-versa. It might surprise them that God is the supreme scientist – and dramatist. God also knows what we are thinking – God doesn't make mistakes. Telepathy is the method of communication in heaven. The writer explains this from a 'religious experience' not scientific deduction. The writer believes that everything is logical and could be understood 'if only we knew how'. The notion that man is 'somehow' linked to God is not unlikely.

Perhaps a brief discussion of our heavenly Father might help to clarify how the writer thinks of God. Firstly, God is omniscient and omnipresent. We cannot be because we are 'single entities' not multiple entities like God. We can only think of one thing at one time. Each entity of God is God, who can act as a single entity with all the knowledge and power necessary for the occasion. Thus, God has an infinite form and a finite form that can be in a million places at once and still be the same God. The finite form can also be incarnate as in Jesus, known also as the Christ (Anointed) as in Jesus Christ. The Holy Spirit is the communicator and executor of the Father and the Son. There can be no disagreement in heaven.

The writer speaks of the Son as the finite presence of the infinite God and can exist or be present in any number of places at the same time. God is truly the Father and fountainhead of all knowledge and power that flows to the Son (his finite form in any shape or style – such as a bright light) appropriately to the occasion. Jesus said, "I and the Father are one" (John 10. 30). He is not saying that he *is* God. He is saying that he has knowledge of, power and authority from, and union with God. Jesus is related to earth. Other incarnations (and revelations) must take place in other civilizations for the same purpose elsewhere in this vast universe.

The writer's purpose is to knit together two usually opposing subjects that eventually combine into one. Religion (or metaphysics) and science seem to be like alpha and omega, separated by a whole alphabet. However, truthfully there is no incompatibility between science and divinity and both complement each other. It may surprise some people but there are substantial arguments, new astronomical evidence and conclusions in support of this proposition.

Admittedly, the general view of religion seems old-fashioned and out of touch with the modern view, but the New Testament is just as true now as it was two thousand years ago because it is designed for a divine society, or to show the way to a divine society. Such interventions are impossible to prove, but that doesn't mean that they don't exist. A divine society requires God to intervene to keep perfect justice. The reader might feel that this does not apply to our society, but that is because we see only the short term, not the long term as God wills. Our earthly society will end and be transformed into a divine society at God's appointed time as prophesied in the Bible, called Judgment Day. Note the difference between long term and short term, or the narrow view and the broader view. Faithless people want to see instant justice but this would thwart God's wise plan and dramatically change the future.

The writer wants to enlighten people about the whole universe, not just part of it. What is religion to most people is metaphysics to the writer who has more personal experience of this than most people and, hopefully, more knowledge, otherwise he would not be writing about it. Metaphysics is, for the writer, the scientific study of the supernatural in relation to Christianity. For instance, the powers and knowledge of Jesus were supernatural in metaphysical terms but miracles in religious terms. In thinking about this the search for evidence for the intervention of God in the daily affairs of mankind has been more positive than negative. In fact, in the long view, it is very positive. In the short view, not nearly so easy to see, and no 'hard' (scientifically acceptable) evidence, except the word of St Paul and the apostles. Personal religious experience depends upon the person – and

God. Contrary to popular scientific belief, knowledge of God has more practical importance than scientific knowledge.

People tend to judge religion by the way religious people behave and think. But we have to take into account the different personalities people have, the quality of their thinking and their feelings about things. What sounds like a simple commandment expressed in a single sentence can be interpreted in action in different ways and different degrees according to all the possible circumstances.

Thus we come to action within the civil law and sometimes without it, as in the issues over evolution theory, homosexuality or abortion. Fortunately, in civilized countries the civil law usually follows what common decent behavior dictates based on reason and compassion (for loved ones and others) as well as ourselves without regard to any religion. True religion should be practised every moment of every day through natural behavior "in favor with God and man". The laws of nature and God determine the consequences. If deeds are offensive to a civilized society it is not unlikely that they are offensive to God, but God can do things that man shouldn't do, because God is perfect but man isn't. Man's judgment is subject to error, especially when the judge, or judges, have a predisposition that warps or corrupts their judgment. Biased courts are notorious for their biased judgments.

Another philosophical idea about purpose in the universe is that life springs up wherever it can. It depends on its adaptability for its survival and development into ever-higher forms. This depends on its ability to anticipate what is likely to happen in both the short and long-term future. In other words, intelligence or understanding at the higher human level. This is where the idea breaks down, or appears to break down. Divine life requires faith as well as understanding to live unselfishly and to possess the blessing of a loving God (peace beyond understanding) in a divinity loving system. It seems to be beyond man except in Jesus whose divine purpose was and is to lift us up to the divine level.

The cosmology of today is out of date too, because it is based on 3-dimensional thinking instead of Einstein's 4, integrating time with space, which also changes the appearance of the universe and the interpretation of the red shift as will shortly be explained. The description of the almost unbelievable shape of the universe is imaginable but not in one picture because of the time factor. The simple explanation of Einstein's problem (the likely collapse of a spherical universe to a point) is that there are no straight lines, and there is no center. Also, there is no absolute motion, only relative motion. All motion is only relative to other objects. There is no absolute reference point from which position can be measured. Scientists agree with this last statement.

The path of light and gravity is circular at a great distance. Therefore the sum of the gravitational forces at any one point in the universe is not the same as straight line forces which apply at relatively short distances between contiguous galaxies forming clusters. This can be understood better by reading chapters 2 and 3 that describe the extraordinary shape of the universe. The shortest flight path in a 4-dimensional sphere must be geodesic as on the surface of earth. Circumnavigation of the universe must follow a line that curves 360 degrees. Halfway along the journey the curvature must be 180 degrees (in the opposite direction) relative to its starting point.

A subjective straight line journey (to the observer in a near light-speed space ship) begins conventionally then slowly, as it approaches about 3 billion light years, is in process of changing direction, like motion on the surface of earth that turns downwards following the curvature of the surface. The journey to a destination at the theoretical limit of observation (the observational boundary) would approach it at 180 degrees to when it started. This prevents the gravitational collapse that Einstein feared but doesn't stop contiguous galaxies at relatively short distances from each other forming clusters.

The subject's importance to the non-scientific general public is that the universe never began and will never end, and that it is a closed, orderly system with religious implications. This is explained later in some detail. The presence of intelligent beings in a totally material universe would be an anomaly (and irresponsible) if there were not an intelligent design and a supreme spiritual designer. Thus, both the universe and God display perfect logic. The writer also explains God's revelations to different cultures and how God works relatively to man and nature. He confirms and supports Jesus, his teaching and his claims. Nature displays perfect logic in physical terms. God displays perfect logic in moral or spiritual terms.

The writer has a predisposition towards scientific or logical thinking as will become evident in his apparently critical or cavalier attitude towards the Old Testament, as Jesus showed in his judgments in the New Testament. The writer's view is that any revelation of God to mankind is relative to the culture and conditions of that time and place. This is a view that requires a lot of thought to be seen to be justified and is discussed later. There might be many revelations but there is only one God. God is omniscient and omnipotent and is the first cause of everything, but he is not quite the kind of God that some people think he is.

It is illogical to form a theory of the universe with living beings that does not include a Supreme Being. If omnipotence is possible, and there is evidence that God is omnipotent, we must allow at least the possibility that there is a God. We must also allow the possibility that such a God is immanent, somehow part of some, or all, life. Such a God could be the source, the continual creator, of all mind and matter. If there was ever a time when nothing existed how could the universe or God come into existence? Time itself has an existence of a kind that we cannot understand, but it is necessary for motion as we understand it (not instantaneous motion that doesn't require time). We need to look at the full picture, not half the picture. The writer approaches the study of God as a perfectly logical being as he sees nature as perfectly logical. Man is 'logical' to his own level, but not perfectly logical. Furthermore, that which is perfectly logical cannot be illogical. Logic also exists in the field of ethics and art as well as science. Every act has a logical consequence. In the moral field some people call it karma (retribution or reward).

There can be good prophets and false prophets but there can only be one perfect prophet, who teaches the truth for a divine (perfect) society. God cannot sin so there must be justification for this policy: bad religions (or worldview) for bad people (such as practise human sacrifice) to lead them even further astray and eventual destruction; other religions to suit other kinds of people, and so on. Think of God as ruling the universe like a mirror reflecting justice but in the long view, not the short view, so common in human society. Salvation to the kingdom of God is everlasting so believers have to be patient. God is very patient. If God worked immediately as some people think he should, there would be no history or evolutionary development, meaning reality, because everything would be accomplished at once. God neither acts nor thinks like human beings think he should. Instead he sent us Jesus Christ. Man cannot attain divinity in this life but he can aspire to it. It is his aspiration and effort that would make him savable.

It is natural to human logic that all revelations of God to all kinds of people should be the same, but Jehovah tested Abraham when he told him to sacrifice his son, and stopped the sacrifice taking place. The sufferings of Job were inflicted by the Devil under the oversight or permission of Jehovah. In both cases the end result was good for all concerned. Such 'testings' are not random but based on reason. By the same logic the writer suggests that all less than divine revelations can be interpreted as testing the goodness (sinlessness) of the people concerned. There is only one religion that is perfect, both in the social sense and the spiritual sense, so much so that no one can obey it perfectly, and that is Christianity. In explanation, if only we were perfect, we would understand why, because we would possess perfect inspiration, divine blessing and knowledge like Jesus.

God may not cause ignorance and suffering directly; these things are the natural consequences of sin, or held to be caused by evil, ignorant people or spirits such as the Devil and demons. People we can see but the Devil etc. can only be believed to exist because the phenomena they cause can usually be explained by natural causes. Jesus took the spiritual view and proved its effectiveness in healing but that still doesn't prove that the Devil actually exists. The phenomena of spirits could be God's or nature's way of effective communication with any human being. As Jesus said, "For whether it is easier to say 'Your sins be forgiven' or say 'Arise, take up your bed and go home'? (Mt. 9. 2-6). What Jesus meant was immediately understood regardless of what expression he used. Perhaps the Holy Spirit in the knowledge and power of God accomplished the deed. The sense of the existence of spirits lies close to the surface in the human mind, especially in the presence of inexplicable phenomena justified by our own awareness of how little we truly know. Unsubstantiated feelings can be very powerful, especially in unfamiliar or 'spooky' surroundings. God also works his unspectacular (impossible to prove) miracles through people, unforeseen events and coincidence.

Cultural values reflect the character and judgement of people according to their reasoning and experience. They are unlikely to change without a change in circumstances or education. Their religion seems designed to match their nature. God is not a Christian evangelist. That is not God's purpose. If it were we would all be Christians and we would not need faith. We would possess the knowledge of God. Jesus is the evangelist but he only draws (or God draws) certain people to him. God does not think like people do who respond to the present in simple ways. He knows the right thing to do to produce the end result in the right (logical) way, even if it seems horribly wrong to humanly reasonable people. Maintaining righteousness and logic in heaven and earth keeps the universe orderly and perfect. God can put everything right as Jesus taught.

God is the Head of the universe

The writer's purpose is to inform and inspire. God is the perfect artist, dramatist, scientist and judge. Creation is a divine art form as well as scientific. It is true that God moves in mysterious ways. "My thoughts are not your thoughts, neither are your ways my ways, says the Lord. For as the heavens are higher than the earth, so are my ways higher than your ways and my thoughts than your thoughts" (Isaiah 55. 8-9). If the writer upsets some people, follow the arguments with an open mind. If you cannot agree with the writer's views just do what Jesus said but do not confuse Christianity with all of the Old Testament (Mosaic Law for instance) as some people do.

Islam is founded on the Old Testament Mosaic Law rather than the new. The writer hopes that Muslims will read this book, understand the logic of the arguments, and take more notice of the New Testament. After all, Jesus is mentioned in the Koran as a prophet who performed miracles, though his teaching and parables seem to have been dropped, including his divinity, presumably because of their unsuitability to the ancient Arabic culture. In modern conditions this may no longer apply, and patience and tolerance are the important considerations between followers of the two religions, otherwise there may be continual conflict in a cosmopolitan society. Let us all remember the rule in evolutionary theory: adapt or perish. There is only one God.

Christian theology appears to be little understood even by followers, especially the Holy Trinity, but the writer has done his best to explain the subject later in the book. The important point for Muslims to understand, if the writer is correct, is that any teaching that opposes Jesus and that results in unchristian behavior is a sin. Muslims are taught that the Bible has been seriously corrupted and that the Koran is the last word of God as reported by Mohammed and recorded and interpreted faithfully. Muslims, or some extremists, take criticism of Islam as blasphemy worthy of the death penalty. By the same logic, if Christians had the same violent attitude the result would be widespread suffering, destruction and death until one side or the other, or both, realized their mutual folly.

God is merciful to the merciful, and unforgiving to the unforgiving in the Christian message. Is not this logical? Allah, as reported from his messenger, the prophet Mohammed, says that he is merciful to those who believe in him and worship him but will cast into hell those who do not believe in him. The Christian description of God emphasises how we treat our neighbor. The Moslem description of Allah emphasises how we treat Allah. Perhaps Arabs respond well to a mighty and fearful God, whereas Christians respond to the idea of love and service, even though it might seem irrational to some people because that is what Jesus taught was the right thing to do, and would be rewarded by God.

God does not need help to do his will. He is omnipotent. Man is inclined to err, but there is no error in God. The writer has met Moslems two of whom he found to be friendly, helpful and enthusiastic, at least as long as the subject of religion did not arise. And if it had, they would have said "God is good"! A group of workers he met were silent, did their work efficiently and left. They may have been new immigrants who could not speak the language.

Short term and long term judgment

If divine punishment followed immediately after sin, nobody would sin. That's the short-term view and the justification for harsh laws. But how could anyone tell the good from the wicked or the faithful from the unfaithful? It doesn't matter to man but it does to God. He wants positive people not insincere people who are only good out of negative fear of punishment. God wants to test us thoroughly so he delays the punishment – giving the sinner enough 'rope to hang himself'. Sometimes the worst sinners repent and become the best Christians, but if they have been executed (short-term judgment) this could never happen.

God follows divine discipline or logic; in him is no error. We cannot follow God because we cannot always understand him. Our whole duty is to follow Jesus because he received the knowledge from God to teach mankind as he said. Thus, Christianity supersedes the Law of Moses. The teaching of Jesus is for a divine society, or to make mankind fit for one. The Law of Moses was for a primitive society. God's Law is for nature or natural law, which is easy to observe but hard to understand because the wicked frequently prosper, and the good die young, but the moral or spiritual law requires deeper insight. Also, the reward of the wicked is of this temporary world, not the next everlasting world, as promised by Jesus.

When a religion becomes established it alters the culture over time just as a culture might alter religion. Then changes wrought over time by a progressive or retrogressive populace might alter both. Several 'civilizations' (organized and powerful) in history have risen then fallen. What causes them to fall? Decadence of the leadership? Too many slaves? They all seem to have been crushed by rising stronger forces. Or is it caused by the natural process of devolution or regression, which is everpresent, overcoming the process of evolution which occurs when there is a need for change and effort in order to survive? Is it happening now with the rise of the east, especially China and Russia, who were retarded for decades by Communism - to the benefit of the west enabling it to forge ahead? There is also a sense of the reawakening and assertiveness of Islam, all of which the world could cope with a hundred years ago, but not so today with scientific advancement in destructive power.

The present economic malaise (disorganization) affecting the USA is now being held to be due mainly to the corruption of bankers. The government will have to strengthen its strategies to return the ship of state to an even keel (see 'Economics' at the end of the book, Chapters 26 & 27). The writer has long been interested in economics because of its importance in people's lives and its effect on a nation's prosperity versus poverty, the result of bad management. Thus, to continue evolving, man requires the will to improve his condition, and competition or fear and suffering to stimulate him. Then he has to feel morally justified in all that he does because this is innate in nature (as in God) and because man is a logical being, but the conscience may be weakened by sin or vice, which corrupts both mind and body, though sinners (all of us) may not be fully aware of this as Jesus was.

This is not to deny the divinity or authority of Jehovah whose revelation (and law) was appropriate to the circumstances of the time, but times have changed, at least in the West. The New Testament is the new revelation designed for the present and the future as the writer later explains in more detail. God's will is always done. God is omniscient and omnipotent by definition, otherwise he would not be God. The writer affirms that this is true, not only in belief but also in fact: nothing happens except it is God's will. What we, or science, or anyone does is all God's will because God understands all things and has a divine plan.

But God's will does not exactly follow people's expectations for various good reasons. God is loving and merciful to people who are loving and merciful, but unforgiving to the unforgiving. God reflects poetic justice rather than mercy. The teaching of Jesus (love God and love your neighbor, even your enemies. Only hate the sin, not the sinner.) confirms this: salvation to the kingdom of God for believers, but descent into a fiery hell for the unrepentant wicked. If God were all forgiving everybody would be saved whether they repented or not. In the writer's experience most people hardly know that they are sinning; they (or we) feel justified but only by their own human standards. No one can stand the judgment of God and survive. Faith in Jesus can save us from judgment. This is God's purpose in sending a Savior.

Many critics have pointed out that an omnipotent God should not allow evil or suffering to exist. The explanation is that all of creation is designed to work logically, but at a level that we cannot comprehend. We live in a real world, not a dream world, most of all a logical world. Pain is the price of reality. If we were not sinners we would not have to suffer. Suffering motivates us to aspire to a higher level of life or consciousness and seems to compensate for our sins – bad for the body but good for the soul or spirit.

Nature treats life as a continuum, from generation to generation. Virtues and vices are passed on and we have to cope with the consequences. How we cope determines our future, whether we are to be 'saved', that is transformed into divine beings in the kingdom of God, or not. Elevation to the kingdom of God represents the final stage in evolutionary development of life where man shows promise of further improvement but needs a miracle, as Jesus promised, to receive salvation.

Metaphysics

A major proposition (for consideration) of the writer is that nature, our whole physical reality, is the 'brain' (physical organ of the mind) or body of God so that all the acts of nature are acts of God. God and nature are one. God is spirit; nature is physical. The unifying factor between God and nature is logic, the tool of intelligence. God is Light (1 John 1. 5) which means understanding. God's mind is perfect in the spiritual realm and nature is perfectly logical (as scientists will agree) in the physical realm, working in perfect harmony with God (as, probably, no one will agree). It would be impossible to prove the nature of ultimate reality if different systems produce the same results. It just seems easier to perform miracles, and for us to believe in them, if our physical reality, including ourselves, is a construction in God's mind. God in Scripture clearly identifies himself with nature, especially in his conversation with Moses.

The miracles and resurrection of Jesus Christ, his appearance to the disciples and his ability to appear and disappear, suggest that all matter is a creation of God's mind, and that Jesus has a special relationship to God. Thus we are all like Jesus in potential: we just have to become sinless and believe. The writer has to admit that all this is a working hypothesis because what else can it be? By thought we can believe or disbelieve, but it really depends on God who is the first cause as Jesus said.

Starting from the premise that God made man in his likeness, think how the human brain works when we are in creative mode. Every thought produces some kind of physical or electrical activity in the brain, which we can crudely observe with the aid of suitable instruments placed on the scalp. When we look at a picture, is the picture reproduced in the brain? How do we know things? How does the brain do all the things it does? It must be very active and very complicated. But so is nature when we look around us, especially the atomic structure. Where is the mind or what is the difference between the brain and the mind? Our will is a function of the mind, so we can deduce that the mind is the higher part of the thought process and in control of the instinctual lower part, or should be in man, but is in God.

Nature (red in tooth and claw) can be cruel as in natural disasters but the Salvation of God through Jesus can make all things right. If you read right through this dissertation you will be, at least, enlightened enough to make an informed decision about your religious beliefs or worldview. Without knowledge or divine inspiration it may be difficult to make the best decision. Belief in God (correct belief as Jesus taught) is an ideal answer to all our problems, and the hope that there is some kind of sensible direction and purpose in life. The teaching of Jesus, which the writer confirms by intuition, reasoning and experience, shows us the way to salvation. Hopefully, the writer's comments will amplify the meaning and implications underlying the teaching, but there is always a wall of sin (self-justification) that seems to prevent us from receiving full enlightenment.

God rewards those who make the effort to seek truth with an open mind. Bending the truth in defence of religion is wasted effort. God doesn't need defending. Other religions might feel justified in enforcing their beliefs if their doctrine supports it, but offence can be counterproductive if it raises justifiable opposition, possibly leading to extreme violence. Religion is often blamed for troubles and wars but the deeper cause is difference of cultures. Warlike cultures will always be at war with somebody as long as some people have serious needs.

Nature is seen to be perfectly logical in the scientific sense. God is perfect in the widest moral sense including righteousness, which is another form of logic; so God, nature and mathematics (also perfect logic) are all logical and without error. Nature can do no wrong and God cannot sin. Either our ancestors or we have sinned causing (our) suffering and death, therefore we need a teacher and a Savior. This is a book that probably no one wants to read but should, because it leads to very important conclusions. Sin seems to have lost its relevance in these days, but that would only apply to temporal life. Sin in divine terms means ungodliness; in scientific terms it means illogical (mistaken), negative, bad or destructive behavior.

When critics confronted Jesus, he answered them. "Do not murmur among yourselves. No one can come to me unless the Father who sent me draws him; and I will raise him up at the last day. It is written in the prophets, 'And they shall all be taught by God' (John 6. 43-45)". Empathizing with another person, seeing from his point of view to understand him, is applied relativity. We cannot see the whole truth from just one point of view either socially or scientifically. It does not change the truth, but it may change our opinion of him. Empathy is a mark of civilized behavior and a necessary form of Christian love: Know your enemies as well as your friends.

Note the words 'I shall raise him up at the last day' (not 'God shall' John 6. 40, also 2. 19). Jesus really has the power of God. He is the finite presence of the infinite God. Except we have faith in Jesus I doubt that we can be saved. To be saved we have to believe in Jesus and obey his teaching, which he received from God. To honor the Son is to honor the Father. 'Faith' is the keyword, because this will save us from judgment (John 5. 21-27). Belonging to the 'body of Christ' (as in the sacrament of Holy Communion) is our key to the kingdom of God. Being 'good' by human standards (or in our own opinion) may not be enough because our sinfulness or state of sin affects our judgment. Even the best of us could not claim to be perfect as our heavenly Father is perfect (Mt 5. 48). The writer has had some religious experiences that have given him some assurance and motivation to write in this vein, described later (Chapters 5 and 23).

In conclusion of this introduction, at the risk of repetition, note that the all-powerful Universal God (the God behind all revelations) is just. He cannot, or will not, save sinners without divine justification. Jesus is our justification. The essence of his teaching, the moral (behavior) and the spiritual (faith), provide the means or the way.

Chapter 2

Science Maps of the universe The 4-dimensional nature of the universe

Science has advanced and demonstrated the complex nature of matter, energy, space and time that changes our perception of a 3-dimensional understandable universe into a pliable concept that changes them in a way that defies explanation with our present knowledge. The writer has extended Einstein's theory of relativity to describe the extraordinary shape of the 4-dimensional universe, complete with two kinds of map that completely changes the current view of an expanding universe. This is because it is now seen that distant galaxies appear to be moving so fast that they will never stop. So much recent evidence conflicts with the expanding universe theory that cosmologists have had to propose a new theory that space is expanding.

It is important for scientists to get the fundamental theory of cosmological space-time right to lead to other discoveries, such as the reason for the finite velocity of light; otherwise there will continue to be a fundamental stumbling block in their way creating all kinds of difficulties. The universe isn't expanding at all, and it never had a beginning. It is regenerating all the time with enormous dust clouds forming new galaxies (discovered by NASA with x-ray equipment in space). It is estimated there is ten times more 'dark matter' hidden in unlit clouds than 'visible' matter. How does nature transform radiation back into matter from whence it came? Matter and radiation are interchangeable. If we add the mass (creating gravity) of all the energy to the mass of all the matter in the universe the sum of gravity should be constant (energy has small mass but a very large volume while matter has large mass but relatively small volume). Energy has no rest mass but it is never at rest. The writer is equating all mass to gravity, which forms the shape and size of the universe.

If there were more mass (and stronger gravity) the universe would be smaller. If there were no mass and no gravity the universe would be spatially infinite and would be 3-dimensional but empty. Einstein came to the same conclusion, but expressed it more elegantly. The universe is so large as it is that it is inconceivable, for pragmatic reasons, that there would be any benefit in making it larger. 'Nature abhors infinities'. Also Nature (or God) encourages variety. The writer supposes that there could be an infinite number of universes in infinite space but they would probably work on the same principles as ours, but what would be the point?

One difference in a 4-dimensional universe is that we can see almost the whole universe from any single point of view, except for a relatively small region of space immediately behind the cosmic horizon up to the theoretical limit of visibility. Einstein's description of a 4-dimensional space-time universe is spherical and boundless. The writer asserts that it is bounded visually by a cosmic horizon beyond which we can see nothing because electromagnetic waves are elongated to infinity to the heat (red) end of the spectrum losing energy until there is nothing left.

The cosmic horizon moves away from the space traveller (like the horizon on earth) as he moves towards it because every observer always sees himself at the center of the universe as we all do, no matter how far we travel. The cosmic horizon is the effect of light waves (a narrow band of electromagnetic waves) lengthened beyond visibility (the notorious red shift) upon which the expanding universe theory is primarily based. It is caused by the curvature of space-time and the consequent rotation of lightwaves through a half circle, 180 degrees, and changing to the opposite direction, increasing with distance. If the space ship continued it would arrive back in the space it left but 20 billion years later in universal (earth's) time-frame – if it travelled at lightspeed.

The current view is that the age of the universe is 13.8 billion years from the 'Big Bang', which started from an unspecified point which can't be identified unless it was earth, but this is based on the misleading red shift. Not surprisingly, considering the nature of space-time, red shifts are now being found denoting distances up to 28 billion light years. There was no beginning and there will never be an end. Like God, the universe is ageless.

Cosmological redshift is seen as due to the expansion of the universe and sufficiently distant light sources (generally more than a few million light years away) show redshifts corresponding to the rate of increase of their distance from Earth (Wikipedia).

The shift in spectral absorption lines towards the red end of the spectrum can possibly be explained by the change in the source of light from the shorter normally invisible wavelengths to longer visible ones. The spectrum itself has moved down displacing the absorption lines upward toward the longer wavelengths. The question arises: why should spectral lines of a particular wavelength not be lengthened to the same degree as the light around it? Recessional motion from the observer should affect all similar wavelengths equally. Is this another 'mystery' of nature? The wavelength should increase to infinity at the theoretical limit of the cosmological horizon, but just before that is reached there should be a minimal amount of radiation that is radiated and picked up on earth as cosmic background radiation from every point in space. This is what cosmologists claim is left over from the big bang.

We cannot make a 2-dimensional map of the 3-dimensional earth without gross distortion, and we can't imagine a 4-dimensional universe in one 3-dimensional picture because of the time factor and the apparently discontinuous nature of our view as we travel towards the cosmic horizon. Imagine a large cluster of galaxies of which we can only see half because of the cosmic horizon. To see the other half we would have to travel towards it until it became fully visible, or look in the opposite direction (from earth) to see the other side of the cosmic horizon. We would travel in any apparently straight line for half a circumnavigational journey, earth would disappear behind us shortly to reappear in front and we would continue to travel without any apparent change in direction to return to earth or the space it occupied when we left.

Our view of space close to our (every observer's) cosmic horizon would always be about 10 billion years old - that is in the universal time frame. The universal time frame is as we observe it on earth while relatively stationary compared to observed light-speed. The rate at which relativistic changes (Einstein's transformations such as mass increase and time contraction, which happen with acceleration) take place relatively to speed is very slow at first, up to 120,000 km per second (40% of the speed of light), then more rapidly to 99% of light speed. Thereafter, as the speed of the accelerating object approaches 100% the mass also increases to infinity, which is impossible to reach, as it would require infinite energy to accelerate an infinite mass. From a practical point of view the ship would have to decelerate to stop to let the passengers get off, which would require as much energy as the acceleration. There is no friction in deep space. The transformations (such as mass increase) are from Einstein's theory of relativity, which have been verified by experiments.

A 2-dimensional map of the universe

The whole circumnavigational journey at (a theoretical) light-speed, to observers on earth, would take 20 billion years. A two dimensional map would be a circle with earth at the center and the cosmic horizon would be the rim. To circumnavigate the universe we would travel in an apparently straight line to any point on the rim then switch to an opposite point on the rim (the two points are contiguous in real space-time) and continue to earth, which would now be in front of us. The actual journey would be continuous. The view of the cosmic horizon from earth (or any single viewpoint) would appear to make two opposite points on the cosmic horizon 20 billion light years apart, but if we travelled there we would find they were in the same place together.

This is because we are looking at images that commenced their long journey up to ten billion years ago. The space-time continuum curves (or folds) into a sphere (note time *and* space) forming our universe. Motion in space-time is comparable to motion on the *surface* of the earth no matter which strait line we choose. All the available evidence supports the writers' theoretical description.

The cosmic horizon would have a small band of darkness where electromagnetic waves are diminished exponentially to zero energy (wavelengths lengthened to infinity). By moving closer, effectively moving the cosmic horizon, we could see what was hidden in that region of space. It is likely that gravity is affected like light to keep the distribution of galactic clusters more or less homogeneous. Every observer from any position always sees himself at the center of the observable universe bounded by a movable cosmic horizon, as we do. Also, the universe looks more or less the same wherever we look except in the very far distance with the most powerful telescope near the cosmic horizon, beyond which we are not able to see anything. This has already been confirmed.

Incidentally, our circular map of the universe would have to be re-drawn to keep us at the center as we progress. If we fold our original map across the center so that our target destination touches its opposite point on the map, no other points will match each other, but if we fold it again to make a quarter circle 2 sets of numbers on the rim will match their opposite points. The writer experimented with a circular piece of paper and marked the edge at one-hour intervals like a 24-hour clock using the numbers 1 to 12. The writer only mentions it in case it has some mathematical significance. Four dimensions must be difficult to represent geometrically.

A global map of the universe

A global map of the universe could be constructed in two halves (two complete maps) with earth at the center of each half. One would need to be a mirror image of the other. A circumnavigational journey over the 'North pole' would show east on the right on the outward part of the journey, then still on the right on the return half. Think of the two earths, one on each side of the sphere, as occupying the same position in space but separated by twenty billion years of curved or folded space-time. We look along curved lines that eventually meet. Both gravity and light follow the same curved lines producing the 'red shift' and counterbalancing the pull of gravity that would otherwise cause the universe to collapse to a point. It also allows galaxies to form clusters but only to a restricted size.

Space travel is similar to travel on the surface on earth, except it takes a lot longer and time passes much faster in the passing outside universe than inside the very fast moving spaceship due to the contraction of time in accelerated objects (another of Einstein's transformations which occur when a mass accelerates or enters a stronger gravitational field). It appears to be travelling in a straight line heading for its target destination, but it is really following an invisibly curved path as on the surface of a sphere such as earth whose surface curves downwards. Time is contracted in the ship exponentially approaching zero as it approaches observed light speed, meaning that it exceeds light speed but only within its own time frame. Light's time frame is zero, meaning that light travels instantaneously (at infinite speed) but only in its own time frame. It only travels at 300,000 km per second to observers in their time frame. It follows that space ships, if they could attain light-speed (which they can't) could travel any distance apparently (to them) instantaneously. This would only be possible if they had no rest mass (the smaller the mass the quicker the acceleration relatively to the finite force).

The contraction of time would be very important in space travel otherwise travelling in space would be virtually impossible due to the immense time it would take. Observers on earth and elsewhere would see the ship travelling at almost light-speed, but to the passengers inside the ship events outside would appear to be happening at a phenomenal speed: planets revolving around other stars very quickly. Observers on planets adjacent to the ship would see the passengers inside the ship as almost stationary. Universal time throughout space would appear to be accelerated to the passengers while their own time would appear to be normal. This is in agreement with Einstein's theory of relativity.

There is also the contraction of space in Einstein's transformations in the line of motion. All the factors together produce the final result. Passengers inside the accelerated space ship would see their true speed in their own time frame. Note that 'acceleration' rather than motion is the keyword for transformations. Motion is reversible but acceleration isn't. Acceleration in relativity theory is equivalent to gravity. It is the gravity of the total mass of matter and energy within the universe that warps space into curved space-time that encloses our universe.

There may be other universes but we can never reach them. The word 'reversible' applied to relativity means that it is not possible to define which of two bodies is the one moving, since there is nothing 'absolute' to measure against, from or to. Motion is only relative to some other body. Actually, universal motion is not recognized by nature, only by kinetic energy. Acceleration, on the other hand, increases mass according to Einstein's transformations to store the energy (E = mc squared), 'c' means observed light speed). Every particle is always at its own center of the universe. It cannot make sense in three dimensions without the aid of 4-dimensional mathematics.

Imagine being there.

It is important to remember that there is a time and space difference between the observer and the cosmic horizon that enables two apparently opposite places on the viewed cosmic horizon to be in the contiguous time and space, as though space-time folds to meet its opposite parts. Note that space-time curves one second for every 300,000 km of space, because the curvature governs the observed speed of light. All is viewed relatively to the observer. The geometry of 4-dimensions is different from 3-dimensions, though in some ways is similar. Images in the mind are 3-dimensional. Only by moving the point of view, in fact or in the imagination, can we see the truth from any chosen place and time. Failure to do this means that we are trying to see things from a distance, which can be quite different from what is really happening. It is a basic rule of relativity. It is also a basic rule for understanding people, seeing things from their point of view.

If scientists cannot understand this they should ask themselves how much does space-time curve per unit of distance. They all seem to agree that space-time curves but no one has speculated by how much. Einstein never made any suggestions, which is not surprising because no one had any idea in those days how big the universe is or how much mass it contains. Also, a 4-dimensional universe does not make 3- dimensional sense, which is how man thinks. It must be a startling or novel idea to people who hear it for the first time, but the writer assures them that, after a while, it starts to seem familiar, even commonplace. Einstein's theory of relativity is difficult to imagine because it is illogical to familiar three-dimensional geometry and thinking, but it also helps to solve many otherwise unsolvable problems.

Motion in Space-time: a diagram

Imagine a vertical circle (or ring standing on a its edge) representing the universe, then an axis at a rightangle to its center representing time (like the x and y axes of a graph). A space ship is moving along the line of the circle making a complete circuit of the universe at light-speed. A point on the circle representing the position in time and space of the space ship moves parallel and equidistant to the time axis as it goes around the circle. As the point moves it makes a spiral line that records position in time and space. One complete revolution would produce a spiral (a single revolution) that is 20 billion years along the time axis and 20 billion light-years around the circle. The actual length of the spiral line would be longer than the circle's circumference because we would have added time (horizontal plane) to space (vertical plane) in the diagram. In reality the line would be the same as the time axis as though time had moved through the ring leaving no evidence in the diagram. Instead of travelling in the space ship we could wait on earth and still meet the space ship when it returned, but we would have to wait for 20 billion years. The ship and its passengers would arrive, in their own time frame, immediately; theoretically, because matter cannot accelerate fully to light speed. Light or any electromagnetic waves (which have no rest mass) transform immediately; there can be no acceleration because that would take infinite time. Matter can never travel as fast as light because it would have to accelerate to infinity, apart from acquiring infinite mass.

The image that we see of an object near the cosmic horizon would be 10 billion years old. It would take a spaceship at least another 10 billion universal years to arrive. Thus, the target destination would be 20 billion years older than it originally appeared in earth's telescope. Einstein called space-time a continuum because time is curved the same as space into a seamless shape that behaves geometrically like the surface of a sphere. Visualizing the universe in a single picture is impossible because a single picture does not incorporate changes over periods of time, or changes in time and space from other points of view. A time-lapse video camera would only show the changing scene over time as we can visualize. What we see now through the most powerful telescopes at large distances is a view of how galaxies looked like millions of years ago in positions that have been grossly distorted by the curvature of the line of sight. The line of sight follows the curvature of a sphere up to halfway around the universe.

If we drew a line of sight from earth to two opposite points on our 2-d map of the universe then folded it so the two ends met, it would represent a true map, but only for that one (joined) line. Everywhere else would be disjointed. It is the difference (or passage) of time in travelling to different places that makes the 4-dimensional universe possible. A three dimensional universe could not be bounded, would be infinite, and gravity, if matter was sufficiently dense would create a closed universe such as we have. Einstein reached the same conclusion when he said that the size of the universe would be inversely proportional to the mean average density of the matter within it (described more fully in Chapter 3).

Cosmic Horizon discovered

In the NZ Herald Sept. 15 2006 it was reported from the journal Nature that University of California astronomers, Richard Bouwens and Garth Illingworth, using the Hubble telescope exploring the limits of observation, found possible evidence for only one galaxy when they expected many. This is what one would expect to see as we look at the cosmic horizon. Only the very brightest quasar x-ray (a little shorter in wavelength than light) galaxies would be visible, then nothing. With the red shift effect high x-ray emitters would become visible (or more visible) as light.

Quasars evidence of regeneration

Quasars are evidently 'young' galaxies at a very active stage in their development whose centers emit huge amounts of x-rays and light. Their existence tends to confirm the constant regeneration of the universe as older galaxies burn up their nuclear fuel. Nature must have a way to convert radiation back to matter to keep the process going, but nobody has found it or suggested a theory. It probably has a connection to black holes, found at the centers of galaxies.

The rotation of light waves moving through curved 4-dimensional space-time causing the red shift could be experimentally verified. In fact, in one respect, it has been verified with observation of the well-known Supernova 1997ff, which shows a large difference between its distance calculated by known brightness compared to the red shift. The writer used this difference to roughly estimate the diameter of the universe. On reflection, the brightness should have been dimmed to the same degree as the red shift unless it began its journey as ultraviolet, x-rays or shorter high-energy gamma rays. The red shift is about the absorption lines (denoting the presence of elements in the light source) moving to the red end of the spectrum as well as the dimming of very distant quasars and supernovas, called 'la supernova', a type of astronomical light source called a 'standard candle' essential for measuring great distances.

If the earth were made of glass

If the earth were made of glass and we could watch the circumnavigation of an aircraft, when it reached the opposite side of the earth it would appear to be flying in the opposite direction relatively to that which it began and it would be upside down. The writer suggests that a similar rotation through the 4-dimensional curved space-time continuum is responsible for the red shift in light from the cosmic horizon. The same should apply to gravity diminishing to zero at the cosmic horizon, to prevent galaxies forming over-large clusters or one universe-wide cluster. Think of a square beam of light being twisted and bent 180 degrees by gravity as it travels across the universe for 10 billion universal years.

Both light and gravity from galaxies on our cosmic horizon would fill space, relatively to them, in all directions. A bright quasar galaxy just this side of the cosmic horizon in our east should send its light (and gravity) to lighten our western sky. Its gravity would, presumably, tend to counteract the pull of gravity of galaxies from the opposite side. This might give a more balanced effect of gravitational forces that might otherwise cause the universe to form one big cluster of galaxies that would collapse that worried Einstein. Both light and gravity would follow a curved path that some mathematician skilled in 4-dimensions could demonstrate.

Gravitation bends light

If a beam of light were sent horizontally across an upward-*accelerating* elevator it would hit the opposite wall a little lower due to the time lag. The path of the beam would appear to be curved. Acceleration and gravity are equivalent in relativity. We feel acceleration just as we feel gravity, except that gravity exerts force equally throughout the body but acceleration causes pressure on the outside, usually from the back of the seat, or rapid deceleration as in a crash. Einstein predicted correctly that a beam of light would curve when subjected to a gravitational field. But in overcoming the gravitation force the light would lose some of its energy, taking a little longer because of its curved path, and its wavelength would elongate showing a loss of energy and a red shift.

Light from the visible edge of the universe would follow a curved path (bent by gravity) up to 180 degrees to reach us. The distribution of matter (galaxies) throughout the universe is approximately the same wherever we look. Galaxies are found in clusters that rotate about a common center. What prevents them from forming ever-bigger clusters? The anti-gravity 'constant' that Einstein originally thought must exist (to prevent gravitational collapse of the universe) actually does: it is the diminishment of gravity (like the diminishment of energy in light – elongation of wavelength) over very large distances where the exponential curvature of space-time becomes significant. Or it could be that gravity extending beyond 'our' cosmic horizon counteracts opposing gravity. Whatever the correct explanation our universe is not collapsing, and it can only be the curvature of space-time that gives it stability.

Chapter 3

Regeneration of the universe

The universe is constantly regenerating. New galaxies are being formed from enormous dust clouds recently observed by NASA x-ray equipment in space (quoted later). An infinite-space would be unmanageable and could not be 4-dimensional because it could not be bounded, and where would space-time curve to? Nothing in relativity theory would work because space would not be 4-dimensional with time as the integrated 4th dimension. It is impossible to have bounded curved space-time and infinite space coexisting in the same place, because the rules of relativity would clash, unless there were numerous curved space-time universes all separated from each other within an endless space.

Einstein said gravity bends or curves space-time and contracts time (clocks go slower). He predicted that gravity bends light rays, which has been confirmed. How much mass (gravity) and energy (which also has mass by virtue of its energy, although it has no rest mass) would the universe need to form a 4-dimensional space-time sphere? Einstein said that the universe is spherical and boundless and its size would be inversely proportional to its mean average density. But he seems to have stopped there because he thought the universe would collapse under its own gravity. He introduced an otherwise unjustified cosmic constant into his mathematics that would prevent that.

The writer suggests that the elongation (and weakening) of electromagnetic waves applies also to gravity, so that galactic clusters do not gather into one big cluster and collapse under its own gravity. Even if this suggestion is not correct gravity works all around, pulling matter closer while more matter farther away is pulling in the opposite direction. Note that geometrically forces may work along lines as on the *surface* of a 3-dimensional sphere in all directions. The effect of this would be to allow stars and planets to form, then galaxies, and then clusters of galaxies. Anything bigger than a critical size, because of the greater distance becomes subject to equal and opposite gravitation. The universe is homogeneous with clusters of galaxies everywhere we look.

Einstein seemed to think of the universe as a sphere, which is understandable, and why he thought it would collapse. But when a space ship travels anywhere it appears to travel as on the surface of a sphere. In other words the perception of the universe changes as we move and the horizon moves. The spaceship has moved to a different time and space, as on the surface of sphere, so the universe, as a whole, looks the same only with different objects around the ship. We move only relatively to the objects around us while we remain, as always, at the center.

People did not know much about the universe at that time and thought it was very much smaller than it is. Hubble's red shift discovery in 1929 indicated that the universe was expanding and that this explained why the universe did not collapse – everybody, including Einstein, thought so and has done since. Einstein then said his anti-gravity constant was the biggest blunder he had ever made, but it wasn't. It was everybody's mistake for not taking into account the effect of the curvature of space-time into a sphere – unimaginable to 3-dimensional-thinking man who tries to imagine the universe in one picture, disregarding the 10 billion years in time difference between the observer and the cosmic horizon, and the unimaginable curvature of space-time into a sphere. It makes nonsense of 3-dimensional geometry. It probably appears highly improbable if not impossible to most people now. Einstein seems to have first envisaged the bending of light and space-time on a small scale, not a universe-wide phenomenon. When he did think of it he saw the universe as a sphere, but then it seemed as though it must collapse under its own gravity. The discovery as was thought then, of an expanding universe, seemed to solve that problem. This may have stopped him seeking a solution to the problem.

The writer suggests that the best description of the universe, experienced by circumnavigation, is the spiral idea described earlier. Any motion is along the time axis as well as the space axis as it would be drawn in a graph. We can't move back in time, only forward. We can move back and forth in space but time moves only forward. We can accelerate universal time by approaching light-speed (in any direction or path returning to earth in some future time) or entering a strong gravitational field, so that a year can pass in the universe but our time (clocks and biological processes, although they would appear normal to us) would be slowed to any degree according to the Lorentz transformation formula. See discussion of Black Holes later.

A geometrical progression to infinity in wavelength in the red shift within a finite distance

Note: Red shift: a shift in lines in the spectrum towards the red, considered to occur because the source of light is receding (dict.) Also, absorption lines denoting the presence of chemical elements in the spectrum of light coming from or through the substance. Doppler Principle: waves become longer when the source is moving away from the observer.

The equator on earth (if the earth were not tilted relatively to the Sun) would receive 100% of the sunlight, then progressively less as we move towards either pole which would receive none (ignoring atmospheric effects). This would be a geometric progression in reduction of sunlight to zero that could be compared to the red shift effect. Supernovas of a certain type emit the same degree or brightness of light at their particular stage of development, which gives us a method of calculating distance by the inverse square law independently of the red shift.

One difference between a three and four-dimensional view of space is that we see almost the whole universe in one view. A 3-dimensional universe would have to extend forever, as it could not be bounded. 4-dimensional space-time can be bent or bounded by gravity as we can see. We know that gravity warps space-time but how much gravity (matter) would warp space into a sphere the size of the universe? Would a super-sensitive gyroscope tilt if it moved in a straight line? Could we measure any change in the wavelength of light over a long distance in a vacuum? Experiments like these could test the effects of curved space-time.

The velocity of light

It is fairly obvious to the writer that the 4-dimensional curvature of space-time controls the observed velocity of light (c), otherwise it would be observed to have an infinite velocity. At present the observed velocity in a vacuum is a constant 300,000 km per second and never changes whether the light source is moving or the measuring equipment.

If this were not so and light were observed to travel instantaneously, our view of the universe would change considerably. Everything would be seen in present time. It would have a different arrangement of galaxies and the observer's position or view would change with motion in a straight line. We could leave the universe. East could be not be seen to occupy space adjacent to West. It seems impossible because we would be trying to mix 3 and 4 dimensional concepts together. Scientists have been trying to explain the 4-dimensional universe with 3-dimension logic and seen, with new discoveries, to be failing.

Chapter 4

How divinity works

As we come to know more about Einstein's 4-dimensional space-time universe and its peculiar rules, which seem to be in contravention to the ordinary rules of 3-dimensional physics, such as the increase in mass with acceleration, the contraction of time and space, the difference between observed time and speed and subjective (view from the accelerated object) observation; the more we might wonder at the 'concrete' reality of everything about us. The universe begins to look like a mind or brain creation of a Supreme Being, whose mind works perfectly just as nature works perfectly. It makes us actors in the drama that we experience as life and death, and everything in between.

This view of a 4-dimensional universe favors belief in an eternal God as the living head or mind of the universe because it is a completely enclosed eternal finite co-ordinate system with self-conscious intelligent beings within it. It is eternal because the future is continually being created unlike space that is finite but whose contents are continually changing but constant in total mass/energy. In other words, a steady-state rational system from which we cannot escape because, to us, there is no other space. We cannot escape from the universe because the boundary moves as we move. There is no absolute motion in space, only relative motion to other objects. With our largest telescope we appear to be at the center of the universe as every other observer sees himself at the center.

The fact that we live in an enclosed universe makes it more personal, more controlled, designed and less likely to be a mindless event. The universe is constantly changing, which is a characteristic desired by the human mind as in changes of fashion, location, styles etc. It has been noted that 'the only constant in the universe is change'. In other words it appears to be a creation by and for an intelligent mind. The writer's observation and experience confirm this. It is further confirmed in the Christian religion as taught by Jesus.

The universe is everlasting, as God is everlasting.

God never began and will never die, so it should be no surprise that the universe, the physical (to us) domain or body of God is also everlasting. Genesis describes God's creation of the world (not the universe) over six days and rested on the seventh. God, through nature, creates all things but not in the literal way revealed to the ancient Jews. How else could he describe it to primitive people who had no knowledge of planets orbiting the sun, gravity and all the other forces and wonders of nature from the very great to the very small? It is for people to discover these things for themselves through the long and arduous process of discovery and evolution.

From the logical point of view, matter/energy can neither be created nor destroyed, it can only be temporarily changed one to the other and back again, a constant recycling. Therefore it can never disappear or die, providing electromagnetism can be converted back into matter. If it couldn't it would have already happened. And what would God have been doing before he created all things? God has the power to perform miracles that defy known scientific laws, so why would he change anything. Nature works perfectly for God. We haven't realized the potential power that God can give us through Jesus to live in harmony with all things in the kingdom of God as Jesus prophesied in the Bible.

It may seem strange to scientists that the writer has such strong views about God and Jesus, but he has good reason, but no evidence to show other people. Belief in a Supreme Being as Head of the universe who is also omniscient and omnipotent could be called metaphysics. A total theory of the universe must also include an explanation of God. The writer treats religion the same way as he treats science: trying to find and understand the truth. The result will be seen to be very controversial. It might upset a lot of people, but it might please others with a similar mind. If people don't want to hear the truth the writer assumes that this is the will of God, because he is omnipotent and never makes mistakes or sins. I know that God knows my every thought and probably everybody else's. The writer has no supernatural gifts or special knowledge.

It is the discovery that we cannot escape from the universe, or move closer to its 'edge', that supports the writer's view that the nature of reality, and the universe, is metaphysical, but it is not a proof. If it is metaphysical it supports the view that God is the Head or Mind of the universe. There can only be one God – unless there are more universes, which can be of no concern to us. Then our main purpose should be to understand the will of God, which we can learn from the only person qualified by his miraculous power, knowledge and resurrection, Jesus.

In the writer's view miracles are possible to the right kind of people - if 'people' is the correct description. Also, they must be acting in the will of God. Science can prove that certain things can be done but it cannot prove that a miracle is impossible, only unlikely in present conditions. It is the writer's view that nearly all miracles (where nature is changed by subtle spiritual forces) are invisible and unprovable in scientific terms. Jesus performed very spectacular miracles two thousand years ago but they (spectacular miracles) seem to have stopped after about the first century, though this is only the impression of the writer. Why is this? We are told to have faith, the essential ingredient in religion and salvation. If Jesus had not performed miracles he would not have been believed. If he had continued in the same manner to the present time we would not require faith. Faith is a test of a virtue required by God. It seems a deliberate policy of heaven to make miracles unprovable.

To emphasize the point, think of the thousands of UFO sightings but no convincing scientifically admissible proof. The writer estimates that it is extremely likely that very advanced beings inhabit many planets just a few lightyears distant. The reason for their visits is to let us know they exist, but in a way that hides their purpose. They are not here to attack us, but to help at the appropriate time; probably the terrifying (as described in the Bible) Judgment Day.

The writer has spoken to a few eyewitnesses, one of whom he knows well, and believes. Witnesses who were forthcoming have often changed because of the widespread scepticism and have become rather unwilling to discuss the subject. God works through nature rather than by spectacular miracles. Jesus worked many miracles so we would believe in him, as he said so himself.

An omnipotent and omniscient God who can and does perform miracles (in defiance of nature's known laws) possesses the universe in a similar way to our imagination being under our control. This would be true in our view whatever the reality.

If a writer or an artist wants to move a mountain in his creation, he just does it according to his will. The difference is that when God moves a mountain it really moves in everybody's perception and experience and stays that way until God or nature (natural forces) changes it. The only way that the writer can explain this is if our reality exists within God's mind. Even if this theory is incorrect, it is a working hypothesis that offers an explanation for miracles within a 'concrete' reality and the salvation of man to the kingdom of God. From the divine point of view it satisfies the requirements of logic or righteousness. From the observer's (man's) point of view, nature is great but God is greater. God is perfect and his works are likewise perfect. Salvation has to be justified. We live in a rational universe.

In man, minds are the product of brains – we are dependent on the healthy working of body and brain. The brain is the physical organ of the mind, so nature is the physical organ of God's mind. We feel more or less in command of our thoughts – anything less would be a sin, but we are all sinners (slaves of sin) to some degree, otherwise we would be divine like Jesus.

Man would be divine if he had a perfect mind because he would then be in communion with God. God is infinite; man is finite. Such men would be 'sons of God'. Jesus was the Son of God, the finite Presence of the infinite God. The knowledge of man is related to his circumstances. His finite state limits his capacity for knowledge. God has no limitations.

Male and female

Spiritually, the writer doubts that there is any difference between male and female. It is unlikely that babies are conceived and born in heaven, so that it is unlikely that there would be two sexes. If the reader finds it difficult to believe that nature obeys God's will, try to see nature in the long and broad view rather than in detail. We tend to live in the present but God's view is forever. God doesn't make mistakes. One of the reasons that we find God (or nature) hard to understand is that God's plan is strategic and long term whereas ours is short term. Long term planning is much more subtle and effective than short term.

We look to religion to find the truth through the revelation, or revelations, of God. It might help the reader to believe if he accepts even the possibility of one or more dimensions beyond our four containing the spiritual realm. Mathematically, any number of extra dimensions can be constructed but it doesn't follow that they can exist in reality. These extra dimensions follow mathematical constrictions; for instance some of them are smaller than an atom. The idea could have some merit because they may be theoretically used to explain the construction of sub-atomic wave-particles with very different properties such as electromagnetism and gravity (ref. Theodor Kaluza, mentioned later). There is much more to the universe than we have probably dreamed of.

If you are a scientist you should read this thesis to gain some knowledge of a Supreme Being hidden behind the physical universe. If you are a believer you should read it to be more enlightened about an advanced science that describes an everlasting universe and how it behaves like the brain or body of God as a total co-ordinate system that is perfectly logical. It will make you a more complete Christian. It will also make the interpretation of the Old Testament more compatible with the New Testament and with science. Believers should welcome science, not be critical of it, nor scientists critical of religion. God is in control of the universe and of science, not just religion. The best religion is a higher form of natural science. To suggest otherwise is to diminish God.

Chapter 5

God reveals himself to the writer

The writer has spent a lifetime researching or thinking how the universe began, also (with a divine push) how God directs the universe. The writer did not discover God by an act of reason, God revealed himself with just a few appropriate words in one short sentence, much to the writer's amazement and joy plus a feeling of fear, no doubt implanted by God. Who am I that God should speak to me? I was contemplating doing something that God did not want me to do. There is no doubt in my mind that it was Almighty God (Our Father in heaven) who spoke to me. What I was contemplating was not achievable anyway, but it was a wrong thought that might have had far-reaching consequences for me. I concluded that God's revelation of his existence was for my benefit. I was about 22 years old at the time.

That is the only time that God has spoken to me. I now seek the 'knowledge of God'; I look for inspiration. Inspiration only comes, as some wit remarked, after perspiration, meaning work or effort. God is always righteous; righteousness is also a form of logic; for instance, reward should follow effort. I have had a few other religious experiences that I have described in appropriate places later for the reader to contemplate. This will help the reader to understand why I can write with confidence on otherwise usually unknowable things.

Christianity

The way that Christianity is often taught in many churches is by presenting the whole Bible as the immutable word of God. The writer has no objection to this as long as congregations accept the teaching of Jesus as paramount and that Jesus is our Lord and Savior, one with God. That is the essence of Christianity. But many people do not accept parts of the Old Testament because of apparent inconsistencies with scientifically proven facts that are explained later. It can also be argued that there are inconsistencies between the Old and the New Testament. For instance, should we still sacrifice animals in worship? What has the sacrifice of animals to do with an enlightened religion? This has been replaced with the self-sacrifice of Jesus on the Cross for our sins that salvation may be, in divine logic, justified. The old has been replaced by the new. This is reasonable and acceptable. There are many parts of the Old Testament as critics point out that are cruel and murderous, but there are many that are wise and good. The Old Testament has been described as a history of the Jews as well as a religious document.

God speaks to Moses Identifies with nature

New knowledge about the age of the earth and its place in the solar system should replace the literal interpretation of

Genesis. This should be acceptable because God created the earth through natural processes. God doesn't say, "Nature did this", he says, "I did it". God identifies himself with nature. For example: God said to Moses, "Who has made man's mouth? Who makes him dumb, or deaf, or seeing or blind? Is it not I, the Lord?" (Exodus 4. 11). This is the closest the writer has seen that God has identified with nature, but he never speaks of acts of nature in the third person as we do. Speaking of nature as acting independently of God would imply that nature has control of the universe, not God.

Scientists think that nature controls natural events (natural cause and effect). The writer believes that nature is subject to God's will and that the logical progression of cause and effect harmonizes with God's will except on those occasions when a miracle is required. This implies that the nature of our reality is the divine mind, which would be perfect like nature and God. We could never tell them apart. The writer prefers a universe with a divine mind than one without – if that were even possible. It is impossible to tell the difference between a mind structure that behaves like reality and a material structure that an omnipotent God can control.

Natural order and intelligible order

If nature could talk she would say she does everything the way she decides, but God has the power and the knowledge to guide her, but he is motivated by reason of the highest order. Nature would not know (or care about) the difference between a divine cause and a natural cause. To explain how this works in a life experience: The writer had been receiving lessons in ballroom dancing with the object of receiving a bronze medal. One evening he found himself in a small ballroom the same shape and size as where he'd practised. He had never danced with this partner before so he thought it would be a good idea to perform the steps he'd practised. The dance went exactly as he planned. Not being skilled in social graces he explained, probably badly, how he had led her through the routine, to which she replied that she had led him. In fact, the routine was designed for the shape of the room. An observer could not tell who was leading because the steps would be the same either way.

Old Testament Interpretation

Nature created the world but nature is the perfect body of God, so God is the first cause. That he did it all in seven of our days is poetry, and completely out of character. They must have been God's days, which would be eons to us. The tree of knowledge of good and evil (the dawning of objectivity, self-awareness, self-will, nakedness and guilt) and the tree of life (knowledge of God) are clearly symbols, not literally trees. Modern, educated people should easily understand this. Aggressively asserting that the account of creation in Genesis is literally true is playing into the hands of critics. The preachers who take this stand think that the whole Bible will lose its divine authority if it is not literally correct, but it is only presented this way because, at the time, it was the only way. God creates everything, preferably by natural means, through nature.

The consequences of preserving the literal view are a widespread lack of respect for literal believers, and a strong undercurrent weakening of the foundations of faith. The evidence that the earth is many millions of years old is so overwhelming that it is futile to argue against it.

The idea that humans are divine in origin might be flattering, but it doesn't seem true either in the moral sense or the physical. We are subject to age-related degeneration and illness the same as animals. We fight for survival when threatened. Our method of procreation and upbringing is virtually the same. Some animals are surprisingly intelligent. Although we are usually inferior to animals in physical strength and acuteness of senses we have some superior gifts: tool-making hands, speech, writing, social relationships, planning etc. It is a miracle that there is a God who cares about us, and sent Jesus to lift us to the divine level.

It is clear that some 'intellectuals' are actively eroding true faith and turning the earthly church into a goodwill society with Jesus as just a talented good man. The end is foreseeable, for without justification of faith, the perceived reality of God and the divinity of Jesus, the church will wither and die for lack of support. This thesis is an intellectual answer to intellectuals' criticisms and questions on an appropriate level that is discussed more fully later. This is a 79,000 word book and covers several subjects, including an assessment of current economics and how to 'save the world' from recurring depressions and their disastrous consequences (Chapters 26 and 27).

We must remember that the Old Testament was written in primitive times for primitive people who believed the earth was flat, if they thought about it at all. They were not ready to receive the literal truth. People of today are mostly not ready to hear the literal truth about 4-dimensional space-time. They cannot understand it but observational evidence and Einstein's mathematics support it.

Chapter 6

Universal God Jehovah and Jews

The Universal God by definition is above all other revelations of himself. All revelations of God are relative or appropriate to the times and the people. Divine revelation that was not relevant would be rejected and useless. The primitive conditions alone would make it impossible to apply objective or perfect divine rule except to improve a bad situation. Divine rule in the best sense is only possible with divine people.

One of the reasons the writer has chosen the title 'Universal God' is because of the past and recent history of Jews. One would think from the Old Testament that the Jewish nation would be the most favored whereas the opposite, because of persecution and the experiences in Germany in the Second World War, appears to be true. Jehovah is the revelation of Jehovah God to the Jewish nation, but the reality appears to be the will of the Universal God. God doesn't need a name because there is only one God who reveals himself in locally suitable ways.

Jews have a reputation for great ability in the sciences gaining a disproportionate number of Nobel Prizes; the most famous one was Einstein. He dropped his Jewish beliefs but still believed in God. The writer believes that Jews are following the old revelation out of misplaced loyalty and hopes that the Jewish nation will find a leader, or leaders, who will guide them into the Christian path and into the "Promised Land". There is only one God, but we are comparing a revelation through Jesus Christ, who proved his divinity, to Moses who was the prophet to a primitive tribe. Remember that right belief only comes after right deeds, and right deeds only follow right belief. Either we try and break the bondage of sin, and may succeed, or we don't try and perish. To break the bondage of sin we only have to believe in Jesus and become one Body. Belief in Jesus is God's test for our acceptance into the kingdom of God.

The only reliable revelation of the Universal God for mankind is through Jesus whom God sanctified and sent for his purpose, to inspire, teach and save. Jesus answered them, "My teaching is not mine, but his who sent me; if any man's will is to do God's will, he shall know if it is from God, or if I am speaking on my own authority. He who speaks on his own authority seeks his own glory; but he who seeks the glory of him who sent him is true, and in him there is no falsehood" (John 7. 16-18). The writer does not claim to have prophetic gifts or insight that is not available to any researcher.

Remember that Jesus was popular with the common people – too popular for the religious authorities. It was the higher priesthood who felt insecure and angry and persecuted Jesus to his death on the Cross. It is always the priesthood who oppose any new teaching because they are the ones most threatened by change. Now is the chance for Jews to re-evaluate their primitive religion before it is too late. Jesus was a Jew by birth, but not by religion. See Isaiah Chapter 53 for a prophetic description of Jesus.

The writer has come to understand that the Universal God is perfectly logical both in the scientific sense and the moral sense. To be illogical would be a sin to God. From this assumption it follows that both God and the universe are logical and orderly. Scientists know that the laws of cause and effect are perfect in nature. What they don't know is that God's moral laws are just as perfect, and that nature and God are working in perfect harmony. God is also perfectly just. God can work miracles but they are always justified. As the body obeys the will of man so nature obeys the will of God. The reason why man can't understand God is because both his logic and area of responsibility is infinitely greater than man's.

The writer confirms the virgin birth, divinity and resurrection of Jesus, the doctrine of the Holy Trinity, the presence of an omniscient and omnipotent God, a 'Day of Judgment', the existence of heaven and coming of the kingdom of God as Jesus taught and promised.

He cannot confirm what he is not sure about and that is the existence of hell, but neither can he deny it. To the writer (who tries to be logical) there is hell on earth for billions of people and has been for thousands of years. Hell after death is stated in the Bible. But what purpose is served in suffering after death unless it is to purify (justify) souls for salvation or peace? In any case, the fear of hell probably keeps many people from sinning more than otherwise. Also, some people might change their minds about hell when death is imminent. Some people get a second chance; while there is life there is hope. The writer recommends the safe course. The beginning of wisdom is the fear of God (Psalm 111. 10). The subject of logic, suffering and divine justification is discussed later.

Divine justification is divine discipline, the maintenance of divine order in heaven and on earth. The writer has had the benefit of religious experiences to confirm his belief, and hopes to reassure all that seek similar assurance, especially Jews who individually show much ability but are held back by a misplaced loyalty to a an older revelation of God, Perhaps the Jewish nation is looking to a leader to lead them to the 'promised land', the kingdom of God. They believe their Messiah is still to come. Were he to appear he would only confirm what Jesus taught.

The benefits of science and the real consequences

Science has also made possible the increase in population of the earth to a degree that seems excessive. Science has progressed but human nature has lagged behind. People are living longer but the frailties of old age are still with us. Science has produced more food for the poorest people of the world, but the end result is that the growth in population has increased the number that subsist in poverty with a high death rate due to malnutrition, disease, warfare or starvation.

The problem of persistent poverty, despite the progress of science or efficiency of production is, quoting from a newspaper article on starvation, 'Too many mouths to feed'. For instance,

Ethiopa's population has increased in twenty-five years from forty million to eighty million. Britain, a relatively rich country, whose population since 1710 has increased from seven million to sixty-one million, who couldn't feed themselves today, can afford to import a third of their food. China recently effectively imposed a limit to families to have only one child, but this has caused, or is seen to cause, problems with a disproportionately ageing population. The population of the earth is expected to increase from 6.8 billion (latest estimate) to 9 billion by 2050.

Options

Unrestrained fecundity causes shortages of food and space that cause further friction. The birth of a baby is usually an occasion of great joy, but if its siblings are already dying of widespread malnutrition where is the joy? If man does not find a remedy for these looming problems then nature will. Animals do not know what their options are, but man does. God has provided man with a conscience and intelligence if he deigns to take notice of the former and use the latter. Perhaps we are closer to animals than we like to think we are!

Last Days

It is evident to the writer that the 'Last Days' of the present world order could occur at any time and it is no coincidence that this thesis has been written. Never has man had such power to destroy himself and less enlightenment to prevent it. Bible prophecy and the natural order are coming together. Therefore, read this thesis if you are not already a committed Christian and make the decision that may save your life. God will not save you if you are not Christian in character, that is to say, if you reject

Jesus and what he taught. Unless you keep heavenly rules in this life you will not be fit for the kingdom of God in the next. God knows your heart better than you know yourself.

The only permanent solution to man's problems is God's Judgment Day and the cleansing of evil from the whole world, saving the elect who will inherit the earth and be transformed and made immortal.

If you are a Christian, reflect on the Christian values in everyday social relationships. God will judge us for what we are, not what we think we are. To be kind is always Christian even if the person does not merit kindness. Our view should be that a sinner is the slave of sin and would not be a sinner if he were not blinded and bound by the forces that shaped him. We should be understanding and empathetic in judgment and deed. We do our best within our ability to prevent present and future evil in the knowledge that Jesus condemned his accusers for their unbelief but he also said, "Father forgive them for they know not what they do (Luke 23. 34).

The style of the Bible is to require belief, whereas the writer has attempted to support belief with reason, or theory based on reason. If it is God's will, it will succeed in its purpose, which is to expand Christianity, to embrace and explain modern (and future) science, and bring many more potential Christians, or lost sheep, into the fold; in other words, make it more believable to more people. Also, using words like empathy to amplify the word 'love' appeals to the writer, and hopefully makes it more meaningful and natural to sensitive people. Sensibility is praised as compared to brutish ignorance which is deplored.

Salvation

Essentially, the Old Testament was a revelation of God for the Jewish nation suitable for the level of scientific knowledge and conditions at that time. The New Testament was a revelation specifically for salvation for individual people, to inspire and prove to them that Jesus was the Son of God through faith in whom they could be saved. Salvation means forgiveness of sin: purification to justify salvation through the life, faith, power and sacrifice of Jesus. It also means the elevation or transformation from the human estate to the divine estate, fit for everlasting life in a future paradise where heaven and earth are joined together. In other words, a heavenly earth where miracles are possible as they were to Jesus. Jesus referred to this as the kingdom of God.

God is Logical The divinity of Jesus.

God cannot sin, which also means that he is never illogical or mistaken; therefore to justify saving sinners there has to be some compensation, a negative (suffering) to balance the positive (gift of salvation). But what can have such significance to balance salvation? The writer believes the answer is in two parts. Firstly, in the life work of Jesus, the Son of God, and in his cruel death on the Cross. Secondly, in man's faith in Jesus that he resurrected himself and has the divine power to save. As Jesus explained to his disciples, the Son has perfect communion with the Father. He does what the Father commands in the knowledge that the Father is always righteous.

Thinking of Jesus as just an ordinary man born of human parents is not good enough. Truly he was born of woman, but his other 'parent' was God who miraculously caused him to be conceived. God's Holy Spirit is the executor and communicator in all of God's acts, so all miracles of God are also the work of the Holy Spirit. Therefore Jesus was both human and divine. He shed his humanity as he matured and became fully divine, presumably at the time of his crucifixion. He said he would return to earth in the kingdom of God. He can take up his body in the time, manner and place of his choosing.

The early presence of physical proof and the later apparent absence of proof of his resurrection put contemporary man in the difficult position of having to decide what to believe. Belief is a gift of God. Follow the right path and you will become a believer. Jesus is the way, the truth and the life: the way to God, the truth about himself and God's gift through Jesus of everlasting life in the kingdom of God (ref. Jn. 14. 6). See also John 10. 17&18: "I lay down my life, that I may take it again. No one takes it from me, but I lay it down of my own accord. I have power to lay it down, and I have power to take it again; this charge I have received from my Father." Preachers frequently say, "God raised Jesus", but that implies that Jesus did not have the power. It is important to show that Jesus has received power, that he is a spiritual being, and can do everything that he claimed. He is the finite presence of the Infinite God whether he is incarnate or spirit. "I and my Father are one" (Jn. 10. 30).

Chapter 7

General information. About the author

If the reader is wondering what professional qualifications I have, the answer is none. I decided that the cleverest people in the world are all failures because they all die, except Jesus. Rather than waste my (little) time learning from others I did my own research, reading Einstein, studying people in relation to Christianity and following science articles in newspapers and magazines. It might be of interest to know that Harvard University has a policy of throwing their new students in the deep end of the pool without teaching them how to swim. They say, "Go to the library and learn what you can there first."

I left high school, in Manchester, England, when I was fourteen. I found that I was ungifted in practical things and didn't fit in very well with other people, so I drifted without pecuniary success. Later I settled down, married and eventually became a 'businessman'. Looking back, I think God guided my life for his purpose, though it was not the life I would have chosen.

Nevertheless, a mixture of major mistakes and minor successes probably gave me the experience I needed though exactly what for I have yet to find out. At the moment of writing (2010) I am eighty-four and starting to feel my age.

If some of my spelling seems American or strange it is because English spelling needs modernizing (notice the 'z'), but I have refrained from introducing too much as it might alienate some readers. I think it is important that spelling should be more phonetic especially because so many foreign (what's the 'g' for?) people (and the 'o') make mistakes when speaking (speeking). 'K' could disappear replaced by 'c'; 's' could replace 'c' for the soft or sibilant sound and 'z' replaces 's' for the deeper vibratory sound (bizy bee). Even Einstein said he couldn't write English because the spelling was "treacherous".

Changes in spelling can lead to unwelcome aesthetic changes in speech, which can become faster and harder to understand (a fault in American teenage females on television who tend to gabble). The national spelling reform societies seem to have solved most of the problems. See 'International Language' at the end of this thesis. Keyboard notation is another antiquated area that needs improvement. Everybody should know how to play a piano or keyboard. I have added a note on this at the very end. Learning to play the piano is a tedious business, but it can be learned in minutes with a simplified system of lines and spaces and graphic representation of notes instead of the present digital system. But to acquire skill and speed takes considerable time. Superior innovative systems require adaptive people to adopt them.

This thesis is entirely Christian in sentiment and fact, and is justified by the principle of truth, which is essential to morality, ethics and science. For Christians and scientists this is a bridge between science and Christianity. Christians appear to need it as well as people who follow science and don't believe in miracles. The writer has mentioned later his own religious experiences, upon which the reader must form his own conclusions.

Scientists should be pleased to read it to learn that the universe has much more potential than they thought and that it is everlasting. At present they think it will run down and die for lack of energy, a well known process called entropy in which heat (the lowest form of energy) spreads itself evenly so there is no potential difference to cause or drive further change.

The relationship between God and nature

To be well informed we need knowledge of both. Knowledge strengthens faith. We all need this knowledge to guide us into right belief, for without right belief or a correct world-view we may follow a wrong path and be lost – in this world and possibly the next. The universe is much bigger and more complex than man has discovered or explained.

Logic of God and Nature

A logical and perfect mind behind reality working in harmony with a logical and perfect universe doesn't seem to have occurred to the general scientific community. They seem to think that the universe follows the laws that man has so far discovered and won't even consider the possibility of the existence of an omniscient Supreme Supernatural Being whose knowledge and judgment is quite different to man's. They seem to associate religion with uneducated people, or wish-fulfilment type of people in a harsh world. The more likely truth is that they don't want to believe even though there are thousands of witnesses. They want scientifically acceptable proof, which is not forthcoming, at least for them, very likely by divine will. Like 'doubting Thomas' they have a disposition to be cynical of anything that does not seem to fit in the general design of the world. Unfortunately, the general design of the world shows the inevitability of age degeneration and death.

Jesus spoke of, and showed, the power of faith. Expressing it negatively, not believing someone is insulting to them. Jesus got angry with such people, not because he felt insulted, but because they would not believe against their own interest. Once again God would be the first cause.

Discovering laws of physics etc, is the way of practical science, but not in human relationships. We are living beings with a sense of spirituality and purpose, and some of us believe the evidence in the Old and New Testaments. Science cannot deny this kind of evidence; it cannot prove that a miracle cannot be performed. Nature accommodates miracles under the right conditions. A miracle is God working his will in defiance of *known* natural laws. Our not knowing a thing does not preclude the truth of its existence. Our knowledge is also very limited.

The tendency among scientists is to assume that because we have found that mathematical and scientific logic seems to explain nature as far as we have discovered that this principle can be extended to explain everything. And so it can, but they have underestimated the infinite capacity of the Supreme Being, of nature and of logic itself, which can be extrapolated ad infinitum to solve every problem. In other words, miracles are possible in the right circumstances. God is omniscient, which man can never be because of the nature of his being, and man can only become divine through Jesus Christ whom God sent for that purpose. The actions and revelations of God have to be understood in the context of the times and the people. What they cannot understand they must believe, or follow their own path without a divine guide.

Three levels of consciousness

There are three levels of consciousness: body, mind and spirit. Body and mind we are well aware of, though we would like to know much more about them, especially how to improve them. The spirit, our attitude to various situations involving ethics, morality, responsibility and belief is the field of greatest contention and importance. This is what Christianity is all about. Our spirit dictates whether we live for eternity in bliss, or die. We can be strong minded and physically capable, but weak in the spiritual department, or any combination of all three.

Strength in spirit is very important if we have the right guidance. It enables us to overcome 'the world' when we are tested with misfortune, as Jesus said, "In me you may have peace. In the world you have tribulation, but be of good cheer, I have overcome the world" (John 16.33). Think of hardship as an obstacle to overcome in the confidence that God and Jesus are with you. Renewal of spirit is renewal of life, and the gift of God through Jesus is life everlasting. Jesus has the power of God; we just need faith. People tend to think that God should act like they think he should, but God obviously has his own far sighted and wise agenda.

God's agenda

God's plan or agenda is to bring the world to the Day of Judgment. This is according to the Bible, Old and New Testaments. There is so much misalignment of minds that it seems a hopeless task to unify all people. The writer's conclusion is that God is in control of everything and we can but play our parts in the scene. Only those who have shown willingness to adhere to divine teaching, surrendering self-will to divine will, shall be chosen. Divine will for man can be found in the words of Jesus in the New Testament, beginning with, "I am the way, the truth and the life".

Importance of the right spirit

A word about the spiritual nature of man: Think about the kind of people who will inherit the kingdom of God as immortal beings. They will be as angels, Jesus said, and there will be no marriage. Presumably, there will be no children because there will be no death. We cannot imagine the kind of life these angelic people will lead because we don't know their capacity or powers. Their code of behavior would have to be truly Christian, as ours ought to be now.

The point the writer is coming to is that Christian behavior is what Jesus taught in order for us to be saved in conjunction with faith. Part of that teaching is to resist temptation, the unclean or ungodly desires of the flesh, the spiritual part that is usually attributed to saints. The writer contends that this is what we should all consider in ordering our lives. We cannot know the rewards, but the writer believes that they can be great. No one but Jesus has ever fully succeeded, but he was and is the Son of God, the finite expression of the infinite Father, sent by God to save all who believe. Our present life should be a preparation for a future eternal life. If we wilfully lust after ungodly things in this life, how can we expect to be chosen for the kingdom of God in the next, where godliness would be the norm and ungodliness condemned? Presumably, with a new mind and body, the saved would have the knowledge and power to be spiritually pure.

Judgmentalism replaced by faith

That is the hard part of Christianity for most people. The easier part, the main part of the Christian message, is to be kind to all people (but not so kind as to help them to do evil) even though we might disapprove of them in some ways. We are not to be judgmental - a common and lethal error. We are quick to see the sins of others but not to see our own. As we judge so may God judge us, whose eye sees all. Some people are born with natural advantages and escape the trials of the less fortunate. Jesus preaches about all these things in the New Testament, often using parables to illustrate the meaning. Christian Salvation displaces judgmentalism with faith. Jesus, who sees as God sees, knows the hearts of men, and if he sees love of God and man then that man will surely be saved, even though he is guilty of sin. We sin all the time, not conscious of much of our sin and how it reveals our character. Pause and think about this from time to time. Enlightenment comes slowly and only when we are ready to receive it, but if we actively seek understanding the process is accelerated. Putting Christianity into practice constantly in our daily life in small things must improve us. Thinking about how we are affecting other people is showing Christian concern and empathy.

Truly Christian women should be 'ladies', and truly Christian men should be 'gentlemen'. All ladies and gentlemen are not Christians, but all Christians are ladies and gentlemen – in a perfect world. It has been said that, in evolutionary terms, women are a million years ahead of men. In social relationships that opinion seems generally justified.

Chapter 8

Telling the truth to an inappropriate audience.

This is a question requiring thoughtful judgment. Parents do not tell the literal truth to their very young children about, for instance, Santa Claus. We have to be careful about saying what we think to violent people. In talking to mentally or emotionally upset or deranged people we speak appropriately, not making their condition worse by aggravating them. In other words, there is no black and white rule in communicating with other people in stressful circumstances. God did not speak of creation in literal terms to the early Jews because they would not have understood what he was describing. He told them the truth, but in a way they could understand. Scientists today think they know truth but it's only the parts they can understand.

The question has arisen: 'Was Peter wrong in denying he knew Jesus at the Crucifixion'? The writer thinks he did the right thing in the circumstances, because it was expected that he would be killed, also he would not have survived to tell the story of Christianity afterwards. He who runs away lives to fight another day. If all the disciples stood by Jesus and died we might never have heard of Jesus. Jesus prophesied to Peter that he would deny him thrice which Peter, under pressure, did exactly as Jesus said (Mt 26. 69 -75).

St John the Baptist was killed because he publicly criticised the King's marriage to his brother's wife, the mother of Salome, a certain way to get oneself killed in those days. Was he right to speak out? Do we tell our enemies where we are going so they can kill us? Situations vary and what is appropriate to one is inappropriate to another. John, described by Jesus as the greatest of mortal men, did what he thought (or knew) was appropriate for him. About Peter, he was a surviving dramatic witness to Jesus' crucifixion. Jesus said, "Behold, I send you out as sheep in the midst of wolves; so be wise as serpents and innocent as doves" (Matthew 10. 16).

The most important factor in this controversial question of truth at any cost is the listeners. Do they deserve to hear the truth or are they the swine before whom we are told not to throw pearls of wisdom? "Give not what is holy unto dogs, neither cast your pearls before swine, lest they trample them under their feet, and turn and rend you. (Mat 7.6)" Don't throw your life away, or your family's life away, if you are questioned by ignorant and dangerous scoundrels. If the truth will hurt someone (especially your children) you are placed in an impossible situation. You make the decision: tell the truth or minimize the damage. More importantly, don't get yourself in that position if you can avoid it. Confessions or statements made under duress are worthless and bring contempt on the perpetrators (as in false statements by prisoners of war). It is hurtful and demeaning to an honest person to lie, but the consequences can be torture and death to tell the truth. The writer feels that there is no definitive answer to this question because it depends on the individual, inspiration and the circumstances.

When we sin, if we are self and God respecting, we experience remorse and try to compensate by doing good, redeeming ourselves before God. Perhaps Peter was a better Christian for having sinned as his later life proved. God works in mysterious ways.

This is a serious question for persecuted Christians facing the prospect of or risk of torture and death for themselves and others including their family. Their captors may give them an ultimatum: leave, or deny Jesus or suffer and die. Consider this: "but whoever denies me before men, I also will deny before my Father who is in heaven." (Mt 10. 33) also, "if we endure, we shall also reign with him; if we deny him, he also will deny us" (2 Timothy 2. 12). Peter denied Jesus at his Cross. Is he a saint or a devil? Will Jesus deny him before his Father in heaven? I think Jesus will save him, rightly so considering his faithful life and Christian works afterwards. We should interpret all laws and injunctions in appropriate contexts considering all the factors. Peter, in the opinion of the writer, was bold, outspoken and impulsive and didn't consider consequences very well. He wouldn't have been in the situation he found himself otherwise.

Questions that are debateable require careful judgment. Jesus said that believers would not come into condemnation but would pass from death to life (Jn 5. 21-24).

While on the subject of Peter, Jesus said to his disciples, "Whom do men say that I the Son of man am?" He got varied answers. "But whom do *you* say that I am?" And Peter answered and said, "You are the Christ, the Son of the living God." Then Jesus blessed him saying, "Flesh and blood has not revealed this to you, but my Father who is in heaven." Then Jesus said, "Peter, upon this rock I will build my Church; and the gates of hell shall not prevail against it. And I will give you the keys of the kingdom of heaven; and whatever you bind on earth shall be bound in heaven; and whatever you loose on earth shall be loosed in heaven." In Mt. 18. 18 Jesus told his disciples that whatever they bound on earth shall be bound in heaven, and what they loose on earth shall be loosed in heaven. In v. 20 he extended it to include "wherever two or three are gathered together in my name there am I in the midst of them."

Later, Jesus described to his disciples how he must go into Jerusalem, and suffer many things of the elders and chief priests and scribes, and be killed, and be raised again the third day. Then Peter began to rebuke him, saying, "Be it far from you, Lord; this shall not happen to you!" Jesus turned and said to Peter, "Get thee behind me Satan; for you savorest not of the things of God, but those that be of men" (Mt. 16. 13-23).

What did Jesus really mean when he said, "upon this rock I will build my Church"? Was it the belief that Jesus is the Christ, the Son of the living God? Or was it Peter the man who holds the key of the kingdom of God? Regarding keys, Jesus, the expression or word of God, is named elsewhere in the Bible. "Fear not; I am the first and the last; I am he that lives and was dead; and behold, I am alive for evermore, Amen; and have the keys of hell and of death." Only God in Jesus would have that awesome knowledge and power (Rev. 1. 17,18).

Abraham

It bothered the writer that God told Abraham to sacrifice his son to test his faith. Abraham built the altar but God stopped him and gave him an animal instead. Thereafter Abraham has been famed for his faith. Why did God give him such a cruel test? The answer came to the writer: because Abraham, by the standards of his day was a great man, but by modern standards he was a cruel man. The story of Abraham, recounted in Genesis, is that he was childless. Sarah, his wife, told him to sleep with her Egyptian servant Hagar, who then gave birth to a son called Ishmael, who became ancestral father to Arabs. Then God told Abraham and Sarah that Sarah would give birth to a child. They were both old and Sarah didn't believe it could happen, but she conceived and gave birth to a son, called Isaac, who became ancestral father to Jews.

Enmity grew between Sarah and Hagar so Sarah told Abraham to send her away with young Ishmael. Abraham sent her into the desert with a supply of food and water to carry on her back. After a while the water ran out and she put the child under a bush and prepared to die. But God created a well. She survived and prospered and God's prophecy about becoming a nation was fulfilled. Perhaps Abraham wasn't being cruel or callous to Hagar and Ishmael, but it seems strange to the writer in modern-day Western society. The purpose of telling the story is to show that God, by whatever name or code of behavior, reflects mirror justice: as we treat others so will God treat us.

In the case of the woman charged with adultery, which of us has not sinned that we have the right, or supposed duty, to throw the first stone? When the woman's accusers finally left, Jesus said to the woman, "Neither do I condemn you. Go and sin no more". The accusers had brought the woman to Jesus to test him because under the Law of Moses she should have been stoned (John 8. 3-11). Scribes and Pharisees frequently tried to make Jesus do something plainly illegal, but each time he answered them in such a way that effectively silenced them. This might be a timely reminder that ancient Old Testament Jewish religious law was political and enforceable by the authorities, as the more extreme Islamic Law is in some countries, though evidently heavily biased in favor of men against women in relationship matters.

Divine Law

In the Sermon on the Mount (Matthew Chs. 5,6 &7) Jesus said "Think not that I have come to abolish the law and the prophets: but to fulfil them. For truly I say to you, till heaven and earth pass away, not an iota, not a dot, will pass from the law until all is accomplished." He was speaking of his many commandments in the sermon which includes the Lord's Prayer, followed by, "For if you forgive men their trespasses, your heavenly Father also will forgive you, but if you do not forgive men their trespasses, neither will your Father forgive your trespasses." The commandments of Jesus are more personal than political and, if practised, would change a whole culture. If heaven protects you, you will be rewarded but more likely in the long term and in the broad view, than in the short and narrow view.

There is another question of motive that I did not intend to discuss, but now I think I should: If the only reason a person chooses to be a Christian is the thought of reward in heaven, it is a selfish reason without much merit. A much better motive would be because we come to agree with the teaching of Jesus. It takes time to develop this higher attitude and it takes faith to believe in heavenly reward. All things work together for God. We are very human, not gods.

Chapter 9

Einstein

If Einstein's four-dimensional theory is new to you just persevere with reading, as it is easy to understand the effects although no one knows how they are produced. Einstein didn't develop his theory of relativity quite far enough, otherwise he might also have accounted for the misleading red shift and why the universe doesn't collapse to a point. He never described the surprising shape of the universe with its cosmic horizon and weird geometry.

The author believes he is the first person to do this. It solves a few mysteries that have baffled scientists in the last few years. However, it is not an easy task to convince people that it is correct because we are dealing with four dimensions of space-time literally which is not apparent in daily life or terrestrial distances, but very important when we look into deep space billions of light-years distant. The writer cannot prove that the rotation of electromagnetic waves through 180 degrees of curved space-time causes the red shift and the cosmic horizon, but it appears extremely likely and fits the available facts.

The curvature of the space-time continuum affects the 'red shift' (in the spectrum of light) from distant stars or quasars exponentially, giving a misleading sign of motion away from the observer. When a light source is moving away from the observer its wavelength increases causing a red shift, which can be used to measure the object's speed, but light's speed is not changed. Microwave speed detectors use the same principle, known as the Doppler Effect.

Scientists haven't understood (or can't believe) the difficult concept of space-time being curved into a boundless but finite sphere, as Einstein's mathematics showed, and its deeper meaning. Three-dimensional space could not have a boundary, unless it was a 'wasteland' beyond the known universe that went on forever. But 4-dimensional space-time can be, and is, bounded like earth's curved surface is bounded to travellers on its surface. They can travel forever but they cannot leave the surface of earth without flying into space. Space travellers can travel forever but they cannot leave our finite space-time. It is bounded by a cosmic horizon beyond which we can't see (there is a progressive loss of visibility in the region) but it doesn't prevent travellers (or matter) circumnavigating the universe because their cosmic horizon moves with them keeping exactly the same distance. The universe is both bounded in that we cannot reach it because it moves, and boundless because we are free to move in all directions but only in curved space within a finite sphere and never approaching the elusive boundary.

Finite space-time

The universe is 4-dimensional in length, width, depth and time. Mathematically, there would be more problems with a 3-dimensional universe than a 4-dimensional one. A three dimensional universe could not be bounded (how could it be?) and it would not have the useful properties of a 4-dimensional one such as the contraction of time to accelerating objects. But we live in a 4-dimensional universe so conjecture does not belong here, except to support the argument against a three dimensional view of space. Scientists generally say that space-time is curved (but very vaguely) in deference to Einstein but they still think and write three dimensionally about the universe while they accept and use the other properties of 4-dimensional space-time as discovered by Einstein.

Expanding space

The latest idea that space is expanding rather than the visible universe is unsatisfactory. If it is expanding, what is its present size and how fast? If it is expanding at the speed of light when will it stop? Is the universe doomed to a lonely existence with an empty sky and no means of regeneration? Essentially, the theory is little different from the present physical expansion theory. The next theory would be that it must somehow reverse and start the 'big crunch'. Also an expanding space universe would be subject to the law of gravity: the less gravity the bigger the universe and the less dense. If the gravity is constant what would cause it to change?

But in the everyday world of terrestrial speeds and distances the differences that four dimensions makes compared to three are virtually undetectable and ignored. Even at half the observed speed of light, 150,000 km a second, very little difference would be evident. The same with space; the effect of four dimensions on observations of stars or quasars (very bright objects) is minimal up to about three billion light years. Thereafter, the effect on the wavelength of light increases it more exponentially to infinity at about 10 billion light years forming a cosmic horizon. Light becomes heat as its wavelength increases losing energy becoming more undetectable.

The 'big bang' theory is based upon a 3-dimensional (conventional) model of the universe that ignores Einstein's four-dimensional description of what appears to be simply space. A cosmic horizon and the curvature of space-time cause havoc with the currently popular view of an expanding universe that began with a big bang. The 4-dimension concept is hard for anyone to believe. It is impossible to imagine in one picture because of the differences in time zones. It does not seem logical to our three dimensional minds. We see space as three dimensional and time as something separate, which is true for practical purposes in our daily lives.

It would be impossible to construct the universe in three dimensions without unlimited space, and the matter in it would either expand forever or collapse to a point. What could limit 3-dimensional space? The way the universe is constructed limits our ability to reach 'outer space'. We cannot even move one centimetre closer to the cosmic horizon.

Velocity of light and curvature of space-time

The time factor distorts a single point of view because different places are in different time zones. Space-time curves one second for every 300,000 km of distance from every observer in all directions. The observed velocity of light ('c') is governed by the curvature of space-time. The writer has never heard of anyone else making this observation. Space and time are literally a composite phenomenon. Light travelling through space records the time change of space but in doing so makes itself appear to have a finite speed.

A fifth dimension

The mathematician Theodor Kaluza thought he had explained the fifth dimension to incorporate electromagnetism (such as light) and wrote to Einstein who encouraged him to publish his ideas, but no further information has appeared. It appears not to be quite acceptable, but has apparently been incorporated into modern inaptly named string theory – they seem more like tiny sub-atomic oscillating flowers with sparkling lights each with its own characteristic circular pattern made of radiant energy, each providing a particular kind of force or effect. It requires ten different miniature dimensions. But it still doesn't satisfy physicists. Can't someone find a better name than 'string theory'? Or a better theory?

Matter/energy indestructible

Do scientists know why kinetic energy increases mass? It is one of the laws of nature that mass/energy can neither be created nor destroyed. The mass increase is to preserve the order in the universe as an object accelerates because there is no absolute motion in the universe, only relative motion to other objects. In the absolute-motion sense every object appears to be always at the center of the universe. So what happens to kinetic energy if absolute motion does not exist? It is stored in increased mass (ref. Einstein's formula: E = mc squared). Acceleration is a change of state as well as a change of relative position. Einstein showed that acceleration is equivalent to gravity. Acceleration increases mass/gravity.

Difficulty of visualisation of four dimensions

The reader will understand why scientists have been unable to understand the implications of Einstein's theory of relativity as applied to the shape of the universe. To our 3-dimensional minds it seems impossible, but nobody denies the validity of Einstein's conclusions. Einstein studied multidimensional algebraic geometry for three years (and "never worked so hard", he said) but it still does not give us a mental picture to understand it in three dimensions. However, we can describe the space traveller's observations as he moves.

Quasars, the brightest galaxies, may be observed to be ten billion light years distant by the red shift calculation, but the light we see has taken (supposedly) ten billion years to reach us. But to get to that distance, the observed quasar, if we all started from the same point (big bang), would probably have been at least another ten billion years travelling even at the speed of light, to produce the image, and that is only on one side of the universe. We appear to be at the center of the universe (really the opposite side and the distance equal to half the circumference). This would suggest that the big bang commenced at least 20 billion years ago (10 billion year-old images arriving from a radius of

10 billion lightyears). The speed of two opposing quasars on each side of the universe between them would exceed the speed of light with the conventional current view, which is supposed to be impossible. Quasars can be observed at great distances all around us, the closest is said to be one and a half billion light years from us.

Because they are the brightest objects detectable they are also the most visible at great distances, therefore they should appear to be more numerous than dimmer galaxies. 500 have been identified. The distances between them appear to be considerable.

Chapter 10

Scientific advancement and nuclear weapons Pakistan

Science is performing apparent miracles now that a hundred years ago were thought impossible. The advance of scientific discovery is proving alarmingly exponential in many fields. Invention is growing faster with the passage of time. The more power we have the more dangerous we become in this unchristian world. The strange thing is that we don't seem able to visualize the direction that advances in scientific knowledge will take.

The prognostication is that man will eventually have the power of God – unless he is stopped by his own evil nature and destroys himself with weapons of mass destruction; or God dramatically intervenes to save life on earth from complete destruction as the Bible prophesies. It doesn't matter whether such an event is 'natural', intervention by UFO's or man-made, all is God's will. The theme of some science fiction stories is that weapons of mass destruction come into the hands of primitive people incapable or unwilling to live cooperatively in peace and they destroy all intelligent life on their planet. When one man or small group dominates a nation or region the risk is multiplied. The same applies to militant intolerant religions with an unwise doctrine and fanatical views.

Nuclear bombs have proliferated and become easier and cheaper to make. Part of the reason for this is the complication of political, economic and other issues, and the smaller nations want the bomb for protection. From my reading of a well-authenticated book, Pakistan in the 1980's had a factory to make several nuclear bombs per year. Apparently, there were complications such as trigger and delivery problems that went beyond the basic principles. India has the bomb too and there is dispute over Kashmir. So far, it is a stand off because no one wants a nuclear war in which everybody loses.

At least, it is a partial deterrent because, from apparently casual comments whose source I can't recall, someone in China seems to think it could emerge from a nuclear conflict without too much damage. China has a type of government that is, apparently, self-elective. It has, in the past been mainly, like Russia, ideologically Communistic but now forced by comparison with the higher standard of living in neighboring countries to change its ideology in the economic aspect and release private enterprise with rapid economic growth and wealth. The result has been economically magnificent.

Only a totalitarian government could make such a comment and survive the concern of its people considering the thousands of nuclear bombs in existence. Nuclear bombs can be made in any size above a critical mass. They can also be made to produce persistent radiation that kills slowly and painfully, and radioactive particles (the 'dirty bomb') that would drift around the world killing everything except insects. The writer has great respect for Chinese and Korean people of whom we have many in New Zealand. They have a reputation for being hard working, talented, friendly and honest. Let us hope that the national leaders wisely reflect those same virtues. Anything less could further destabilize an already unstable world.

They determine national policy at home and internationally. The quality of the leadership determines the future of the country. Will it be peaceful, friendly and prosperous, or will it be self-centered, militant and probably disastrous, both for the Chinese people, for the nation and the rest of the world? Very powerful nations, which China is already, with potential to increase in power, should have a sense of international as well as national responsibility and harmony. Wise government is essential to preserve life and liberty. Wise people would support a wise government. Science has provided ever more efficient means of production that should produce ample wealth and leisure for everybody, removing much of the incentive to make war.

The cost of the uranium ore from which uranium is extracted is less than 1% of the final cost of the energy obtained from its use in electricity generation plants. Naturally occurring uranium contains about 0.07% of uranium 235 necessary in nuclear bombs. For this, and for electricity generation, enrichment is necessary, though to a lower degree for electricity. Enrichment means separation of U235 from the slightly heavier U238 of ordinary uranium in hundreds of very high-speed centrifugal chambers or some other processes. The good news is that electricity can be produced relatively cleanly and there is plenty of ore available. The bad news is that bombs can also be made relatively cheaply. The waste should be a minor problem from nuclear generators if handled responsibly. A very high level of competence and responsibility is necessary in the design, placement and operation of nuclear generators. It is believed that much of the heat in the center of the earth comes from natural radioactivity, from the gradual breakdown of the more unstable radioactive materials transforming into more stable elements.

Hydrogen bombs

These are nuclear fusion bombs that are more 'efficient' than fission bombs but a bit more complicated to operate. The bomb yield is much higher. About 0.1% of the nuclear fuel is converted to energy in fission (breaking down) and 0.3 % in fusion (building up) bombs. The first true hydrogen bomb test was in 1952. It exploded with a force 450 times more powerful than the fission bomb dropped on Nagasaki in WWII. A fusion (hydrogen) bomb can theoretically be made 10 thousand times more powerful than the Nagasaki one. At least they are (apparently) a 'cleaner' (small or smaller) radiation danger. Fusion of hydrogen into helium (with a slight loss of mass) keeps the sun burning, which will, in a few billion years, burn out of fuel. Scientists are trying to discover how to control hydrogen fusion to make a power generator.

About the Chernobyl nuclear reactor fire in Russia (1986), "U.S. newspapers carried daily front-page stories about the radioactivity deposited on our soil by its fallout. The typical amount of radioactivity measured was ten picocuries per liter of rainwater. The one millicurie Joe and I each considered safe to drink [at a 'notorious' physics lecture] was a hundred million times stronger than the ten picocuries that apparently scared the American public in 1986 sufficiently to cause many citizens to cancel long-held European travel reservations." (Luis W. Alvarez, Nobel Prize in 1968 for his studies of sub-atomic particles; World Treasury of Physics, Astronomy and Mathematics p. 730).

Let us hope that fear of consequences prevents future wars. A present region in dispute appears to be Kashmir between India and Pakistan. "It will only be a matter of time before the rising tide of Sunni extremism and the fast flowing current of nuclear exports find common cause and realize their apocalyptic intent" (From, 'Deception' published by Atlantic Books 1977). A substantial part of this book is about Dr A. Q. Kahn. Only desperation or an unbalanced mind could cause a nuclear war, but a nation (or nations) facing annihilation could become desperate enough, especially ones that believe they will go to heaven for killing and have the political power to put their beliefs into effect.

The famous, or infamous, depending on the reader's point of view, Pakistani A.Q.Khan, interpreter with a degree in metallurgy, was almost the sole person responsible for Pakistan developing its bomb-making capacity. He acted in the belief that it is disgraceful that others should have nuclear capability but not Islam. His nuclear activities are available on the Internet, but they omit the part about his being bi-lingual and his interpreter duties in Europe in the sensitive area of nuclear technology around 1979. Pakistan is a politically unstable country and forced to harbor militant and ruthless religious extremist terrorists on its borders and beyond who are alleged to train children and young adults to think like themselves. Hopefully, the adverse worldwide publicity terrorism has evoked has made them reconsider the wisdom of their policy. In early January 2011 Salman Taseer, a member of the ruling Pakistan People's Party, Punjab's progressive governor was murdered by one of his security guards because of his denunciation of Islamic blasphemy laws.

European governments are extending the friendly hand to Muslim immigrants, but their indigenous populations are uneasy about the risks and the growing presence of what is to them an oppressive alien culture being thrust upon them. I doubt whether the government realizes the seriousness of the differences between the two cultures apart from the religious differences, which can make them incompatible.

Egypt: As reported in Time magazine, (dated Jan 17 2011) an explosion outside a Coptic Church killed at least 23. Coptic Christians who form about 10% of the Muslim majority say the government does little to protect them from religious violence and discrimination.

Bombs do not discriminate, especially big ones. Many countries in the world are governed by dictators or dictatorships of small groups, usually backed by the military and police – who depend for their existence on the leadership. They have to answer only to themselves and their motives seem more nationalistic for self-glorification than for the welfare of their people. The people can starve while their country's military grows strong. The suppression of constructive or implied criticism encourages the majority group to become extremist and tyrannical that brings its own inevitable consequences. It is said that power tends to corrupt and that absolute power corrupts absolutely. Such a closed country, unless it is already enlightened, cannot become enlightened.

President of Turkey 1923-1938 Mustafa Kemal Ataturk

From the Internet: His achievements in Turkey are an enduring monument to Atatürk. The world honors his memory as a foremost peacemaker who upheld the principles of humanism and the vision of humanity. Tributes have been offered to him through the decades by such world statesmen as Lloyd George, Churchill, Roosevelt, Nehru, de Gaulle, Adenauer, Bourguiba, Nasser, Kennedy, and countless others. ***"I look to the world with an open heart full of pure feelings and friendship"***. The honorary title Ataturk, meaning Father of the Turks, was given to him by the Grand National Assembly in 1934. He was born Mustafa Kemal.

Franklin D Roosevelt quoted the Foreign Affairs Minister of the Soviet Union, Maxim Litvinov: "Litvinov told me that the most valuable and interesting leader in the world does not live in Europe but beyond the Straits in Ankara and that he was the President of the Turkish Republic, Mustafa Kemal." Ataturk made many peace treaties with surrounding states, even traditional enemy Greece whose leaders praised him.

"My people are going to learn the principles of democracy the dictates of truth and the teachings of science. Superstition must go. Let them worship as they will, every man can follow his own conscience provided it does not interfere with sane reason or bid him act against the liberty of his fellow men." Ataturk fiercely rejected interference of religious representatives in politics, throwing one of them out of his office in a temper. Turkey is a Muslim country but the majority still wants religion kept separate from the state, which was one of his great achievements. He introduced proper education (not religious rote) to the highest level for girls as well as boys.

Ataturk was a brilliant military commander and independently minded scholar, wrote several books and wished to bring his country to be the equal of the best of western developed countries. He even dressed in European style. As a President he saw Islam as a force backed by landlords and feudal leaders against public education and enlightenment, but he only took away their political power and the enforcement of dreaded Sharia or Islamic law.

There was one serious death threat against him but it failed. The greatest threat to Muslims is not violence, but criticism, which is why they are so sensitive about it with death threats to perceived enemies. Criticism should be met with reason in a civilized way. In the eyes of the west their greatest fault is extremism. But how can the west know which Muslims are anti-social extremists as in Afghanistan when they merge with the moderates who fear them? As a religion Islam competes with Christianity and may succeed in numbers in receptive regions, but where Christianity is understood, it will be rejected.

If the leaders of non-Muslim nations take sides (against perceived mutual enemies) and there is to be a nuclear war of mutual destruction God has promised to 'shorten those days for the sake of the elect lest all life be destroyed ' (Matthew 24. 22). The writer mentions 'leaders' because they hold the executive power, often in countries where they hardly represent the will of the populace who only hear what their political masters want them to hear. 'The first casualty of war is truth'.

From one of Ataturk's speeches: "Human kind is made up of two sexes, women and men. Is it possible that a mass is improved by the improvement of only one part and the other part is ignored? Is it possible that if half of a mass is tied to earth with chains that the other half can soar into skies?" In 1919 he led his nation to full independence. He put an end to the antiquated Ottoman dynasty whose rule had lasted more than six centuries - and created the Republic of Turkey in 1923. As President for 15 years, until his death in 1938, he introduced a broad range of swift and sweeping reforms.

One of his reforms was the successful introduction of Roman letters in writing replacing the old Arabic script. Literacy in Turkey improved from 10% to over 70% with great public cooperation in two years. He banned polygamy. He was widely accepted as a model for other leaders. He avoided war if at all possible as a policy and was a friend to his neighbours. He must have shared compromises with them to be so successful a negotiator. He was a most successful military commander (he was sickened by the death of all brave soldiers) but he chose the path of peace and development for his country.

Chapter 11

How God judges Jesus and the beatitudes The Trinity False Prophets

The two great commandments of Jesus are: Love God and love your neighbor. And the golden rule: Treat other people as you would like other people to treat you. God is just: as you judge others so will God judge you. God is merciful to those who are merciful. The way to win God's approval is by doing good and being good, not by empty religious ceremony which only makes yourself feel good, thinking that is the way to heavenly reward. God is not vain. He cannot be deceived by worthless flattery. True worship should spiritually uplift people not incite them to destroy. There is only one God by whatever name he is known. He is selfless. In him is neither sin nor error. His justice is perfect. He repays sin with retribution and with death. In the opinion of the writer the **Universal God**, who is (by definition) the real God behind divine local revelations, watches over us, and all things will happen only according to his perfect will.

From Mt 9. 13, Jesus said to the priests, go and learn what this means (quoting Scripture), "I desire mercy, not sacrifice." The meaning of this is that sacrifice is expensive or valuable and is therefore a great loss to the offerer as an act of worship, yet God prefers to see an act of mercy that costs nothing, but elevates man's soul. This is what truly pleases God. Thus we have a higher revelation of God and a lower revelation to accommodate man who understands sacrifice but not mercy. The message is that all of man's aspirations should follow the higher form if he wants to draw closer to God.

The primitive idea of a tribal God is that he will respond to the needs of his people in war and peace. In other words, self interest. Intuitively, they propitiated him with sacrifices and ceremonies. Sometimes the sacrifices were horrendous. There had to be some relationship between the greatness of the sacrifice and the greatness of a particular need. Morality did not enter into the transaction. Jesus introduced a higher concept of God who gave moral (Christian) rules to man, and promised his followers elevation to divinity and eternal life in the kingdom of God. What is required in Christianity is as high a level of faith and conduct in keeping with the teaching as we are able.

If Jesus was an important prophet, acknowledged in the Koran, where are his teachings, his beautiful parables, his miracles and warnings and his claims to divinity? Why are Christians regarded as enemies of Islam, and Christian missionaries furiously rejected in Islamic countries? All people have much to learn from Jesus. He said, "My sheep hear my voice, and I know them and they follow me; and I give them eternal life, and they shall never perish…I and the Father are one" (John

10. 27-30). Every word that Jesus spoke, and every miracle he performed, confirms his oneness with God.

What we cannot understand we must not reject because we cannot hope to fully understand how God works, rather we should learn more and have a little faith then God will enlighten us. Men love violence as a quick way to work their will, but God loves mercy, and God *is* Light. Without light we are without God. Jesus is the Light of the world (Jn 8. 12). Men may seem intelligent but they may be lacking in judgment. Let Jesus be our guide.

Also, from the beatitudes at the beginning of the Sermon on the Mount by Jesus in Matthew, chapter 5: "Blessed are the meek for they shall inherit the earth. Blessed are the merciful for they shall obtain mercy. Blessed are the peacemakers for they shall be called the sons of God. Blessed are those who hunger and thirst for righteousness, for they shall be satisfied.

In this modern world we have to learn to live together in peace. What is truly different between Christians and Muslims, especially in modern times, are their different cultures and values apart from some of their religious beliefs. In the writer's opinion cultural attitudes and religious attitudes in a region complement each other but that only means they are appropriate for that place and time. To survive, especially in the present time, we have to grow and improve, which we can only do by learning from Jesus as much as we can. War is the worldly solution to worldly problems. True religion removes the many causes of wars.

Hopefully, the day has long gone when common people are mostly illiterate and uneducated. The more enlightened people are the more civilized and humane they become, and the more they tend to agree, at least on principles. When dealing with an important subject, get the facts from the source if you can, or get a variety of views. The more knowledge you possess the more likely you will make a wise choice.

The oldest Koranic fragments from Mohammed's dictation to his followers are dated around 725 AD – a century after they were first recorded. Recommended translators of the Koran from the Internet are: Ahmed Ali, Pickthal, Noble by Muhsin Khan, Yusuf Ali and Shakir.

The Trinitarian nature of God

The Trinitarian nature of God does not mean there are three Gods, no more than man is made of three men. Man has a 'self' or 'ego' that he thinks of as 'me' or 'I'. But he could not function at all without a 'father' (the subconscious part of his brain) that stores and supplies all his knowledge, language and skills as he needs them. The Holy Spirit, who is one substance (in man the nerves that the body needs to coordinate the body and communicate with the brain, and of which all the brain is made) with the Father and the Son, completes the Trinitarian description of God – who says in Genesis that he made man in his likeness. Power, knowledge and gifts come by the Holy Spirit who is the communicator and executor from the Father and the Son. Whoever says that Christians worship three gods when he knows that they worship only one God is a deceiver with an ulterior motive. God the Father has the ultimate power and knowledge and is the first cause of everything as Jesus taught. The human being has mind, body and nervous system, when any part gets sick the whole body suffers and when one dies they all die. God is perfect in all that he is and does. If he allows people to believe in another kind of God, it would be to test their purity of soul and heart, for the pure cannot be deceived. There is no division in heaven. The Devil would be too clever to even think of opposing God.

An infinite God when he appears in one present time and place ie. an incarnation, has necessarily the form of a finite being with apparent, but not real, limitations. This would describe Jesus who could call himself the Son of God and claim 'oneness' with God from whom he received power and knowledge appropriate to the circumstances and his purpose. God is the supreme artist and dramatist and performs his works according to his will. If the writer is correct God holds the whole universe in his mind and we are but his thoughts. We see the order and discipline in nature but this is a reflection of how God prefers to work. What might appear to be chaos to us is all part of God's plan.

The Devil is a personification of the tempter whether he exists or not. The existence of temptation is natural, but the personification helps us to recognize the difference between godly desire and ungodly desire. A personification makes a spiritual or mental construction more intelligible and manageable. An example occurs in the New Testament in Matthew 8. 28 when Jesus drove demons out of two demoniacs and, at their request, Jesus said, "Go". And the demons went into the swine and the whole herd rushed down a steep bank into the sea and perished in the waters. A literal explanation of the event would probably need a detailed psychological or neurological explanation and more. We don't hear much about demons these days but people still behave as though they are still active. If the theory fits the facts follow the theory until a better theory becomes evident.

The most evil (ungodly) deeds are done for selfish reward, or to satisfy a vice or for the pleasure and sense of power and, sometimes, a sense of achievement from inflicting pain and suffering including death. This is regarded in the modern west as psychotic behavior (mental sickness) but this is no help to the victims. Such people usually have a long record of wantonly cruel behavior. The less lethal syndrome is widespread involving mostly hurtful speech but words can cause real suffering. What is missing in these people is empathy and proper sensitivity; in the young it would be 'immaturity' or lack of knowledge and experience. We should not judge but wait for people to judge, because how they judge indicates the kind of people they are. Mt. 7. 1: Judge not, that you be not judged.

Empathy is knowing and feeling from the other person's point of view. This is fundamental to Christianity and to all enlightened social behavior. It is also the basis in relativity in physics to see the truth. Even religions, or supposedly divine revelations, seem to take relativity into account. It doesn't change the reality but it makes it more understandable.

Where is Christian love in all this? Christ is the Prince of Peace. God is Almighty and rarely does as mere man thinks he should. Man should follow Christ, who knows, and was sent to teach God's will for man. Following God is beyond man's capability. God is infinitely wise and powerful. He has his own agents, natural and human, and ways to do his will. We see some of the power of nature, and nature obeys God, but there is much that we don't understand. Or God has designed the world as it is and mankind will destroy itself in God's own time.

The wrath of God is shown in countries where people destroy each other. The leadership is corrupt; wealth is not shared; people are angry at being ignored and cheated and so they rebel. The more potential wealth in the form of oil or minerals etc. is discovered the more (in biblical terms) the house is divided against itself. It is clear that such uncivilised countries cannot govern themselves. If it isn't money it's politics. If it isn't politics it's race. If it isn't race it's religion. Good government requires wisdom and power – and good people to support good government.

Mt. 7. 15-20: "Beware of false prophets, who come to you in sheep's clothing but inwardly they are ravenous wolves. You will know them by their fruits. Are grapes gathered from thorns, or figs from thistles? So, every sound tree bears good fruit, but the bad tree bears evil fruit. A sound tree cannot bear evil fruit, nor can a bad tree bear good fruit. Every tree that does not bear good fruit is cut down and thrown into the fire." This would also be true about everybody, not just prophets.

How could God allow such deception? An omnipotent God controls the universe with a plan that involves all the drama and time of natural evolution, which is designed to produce the desired result in the right way, not short-sighted man's way. God is justified whether, to man, he appears to be or not. If false prophets arise, they are to feed sinners with appropriate food to condemn them and mislead them even more using their own will to punish them. In the overall view they are working to separate the good from the bad. Is not this God's plan? To bring all people, the children of God and the ungodly children of the world to the final judgment on the Last Day as prophesied in the Bible? God cannot sin nor be in error. What appears to be obvious to the worldly point of view is sometimes opposite to the divine view.

Who are the children of God? Everyone should read the New Testament, especially the words that Jesus spoke. He spoke clearly and consistently, and worked many miracles of healing to show that he had divine power and the blessing of God, speaking by the Holy Spirit with whom he shared oneness with God his Father.

Good people know good people, recognizing in others what is in themselves. 'Good' meaning a natural or innate knowledge of God, not just what they have been taught. It is called 'humanity', being humane. In the experience of the writer the most obvious sin of 'superficially religious' people is judgmentalism (meaning 'apt to pass judgment without sufficient knowledge or understanding') tending to cause unjustified rejection of other, perhaps worthy, people.

It is actually religious snobbery similar to intellectual and social class snobbery. Jesus highlighted religious judgmentalism (and Christian duty to our neighbors) in the parable of the Good Samaritan (Luke 10. 29-37). In those days Samaritans were despised by the high-class priests, but they bypassed the man who had been badly beaten by robbers and left on the roadside while a passing Samaritan went to considerable trouble and expense to help him.

God is the head of nature and needs to be fair in his dealings with his creation, considering that he rarely reveals himself. This might explain why he acts impartially to everybody. For instance, it is possible that witchdoctors may perform miracles, though they seem more like hypnotism or 'suggestion' to the writer. All kinds of things can happen apparently against the laws of nature but they all fail sooner or later if they go against God who is the ultimate Judge. "Unless the Lord builds the house, those who build it labor in vain" (Psalm 127. 1).

Design in nature

If religion seems irrelevant to science think about how everything comes to exist. How did the universe begin and how will it end? How did it come to be so orderly? The existence of the whole of creation implies an intelligent creator, which is a question for both science and religion. The writer answers these questions and quotes evidence to support the theory. Mankind does not really understand God. The Bible helps but it needs explanation in these scientific times.

The popular view of Christianity is that it is not tough enough to deal with the wickedness in the world. The warnings in the Bible are that if we are not saved then the alternative is hell. But the popular view is that there is no afterlife and that death is just the end of life. The reality appears to be that nature treats life as a continuum, that nature rewards the descendants of virtuous people and vice-versa, but only in keeping with the life-style of the ancestors. There are many virtues that reward us in many ways but the way of godly virtues that leads to salvation is narrow and straight.

Chapter 12

Size of the universe Mathematics Correlations Mind or matter?

It takes light 100,000 years to cross our own Milky Way galaxy that contains about one hundred thousand million stars like our sun, and it is only one of many millions more galaxies. Nearly all the stars we see with the naked eye are in our spiral arm in our own galaxy (the Milky Way that is seen running across the sky).

The time dimension is different from the spatial dimensions in that it is being continually created, which is why it can be infinite. We can examine the past but not the future because it doesn't exist yet. We can examine the evidence from the past in the present providing it has survived but we cannot travel to the past, which is fortunate otherwise we might change it and destroy our own future as in science fiction stories. They get over this difficulty by creating multiple universes.

It may not be of scientific interest but this alters the writer's feeling about the universe. The sense of being within our own private space-time instead of unlimited space (although it is so huge that it might as well be) confirms the existence of God, and the likelihood of the universe being mind than matter. The two ideas go together because mind implies a knowing being, God. If the universe appears to be controlled by a perfect mind, it might as well be for all practical purposes. In which case it is in everyone's interest to learn everything they can about God, on whom we are completely dependent. We have the example of Jesus, born part human part divine, how he lived, overcame temptation, fasted and prayed and became wholly divine, one with God. If Jesus could talk immediately he was born, as recorded in the Koran, he would have been world-famous, even in those days.

The idea of a finite universe is really more congenial than an unlimited or infinite one. Nature is said to abhor infinities. For one thing it is more manageable and anything could get lost in infinite space. Infinities are very unpopular with mathematicians because equations producing infinities are usually useless. Nothing can be lost from our 4-dimensional universe, not even a speck of dust.

For those who might be interested, E = energy; m = mass; c = velocity of light, which is a constant, so the relationship or ratio between mass and energy is also constant. When mass changes (with acceleration) so does the energy stored in the same proportion. The velocity of light in a vacuum is always 300,000 km per second. Kinetic energy is energy of motion. 'mc squared' means m times c times c. The increase in mass from acceleration is exponential and is extremely small until its velocity reaches fairly close to the velocity of light (which is why we don't notice it at terrestrial speeds) when it increases much more rapidly as it *approaches* the velocity of light. It can never reach light velocity because the mass would approach infinity and require infinite energy to increase its velocity. A more exact figure for light velocity in a vacuum or 'free space' is 299,792.458 km per second. It always measures the same speed regardless of motion of the light source or the measurer, which is why it is called a 'constant'.

The speed of light should be found to vary slightly relatively to force of gravity and curvature of space-time in different places. The sum of gravities in the universe is responsible for the curvature of space-time, which is responsible for the finite speed of light. No gravity = no curvature = no limitation of light-speed. When we measure the speed of light we are dependent on distance (space) and time. Gravity (and acceleration that increases mass and gravity) contracts time and curves space.

Mathematics, art and correlations

With the aid of very sophisticated analytic (algebraic) geometry, necessary for higher dimensions (mathematics increases the power of the human mind to solve problems) Einstein was able to describe a seemingly magical side to our universe which no one had ever suspected. The writer believes that logic, extrapolated to infinity, is the fundamental principle of everything in the universe. There is no end to its potential complexity. Everything can be explained and there is no problem that cannot be solved.

Perhaps the words 'symmetry' and 'harmony' as well as 'logic' convey the meaning better, but it is the correlations that are at the heart of recognition of art or design by the intelligent and responsive mind. Mathematics is the supreme example of correlations. Everything has a meaningful relationship to everything else. Logic is perfectly correlated as in mathematics and is integral to nature as it is to a perfect mind. Anger has a correlation to cause. Rest has a correlation to effort etc. The difference between the different levels of art is that the correlations in the lower levels are obvious while in the higher levels they are more subtle, complicated or hidden. The style of correlations might be between the work and its author that might impress like-minded people. We don't have to understand the correlations, we only have to like the work; our subconscious does the 'seeing and appreciating' then transfers the appropriate emotion to the conscious ego.

There is good art through to evil art. High-minded people appreciate good art, and lower-minded people appreciate lower art. It is called 'taste'. Lower art can be very clever in construction, which is appreciated by those with lower taste, but regarded by people with better taste as clever but distasteful. Art has character, especially heard in music. For instance it can be light and bright, or the opposite and depressing or angry and menacing. The works of mentally sick people can express their moods. We can't consciously change our natural taste but we can, over time, elevate our level of thought and behavior if we are motivated – or the opposite.

About God as the perfect mind. Is the universe mind or matter?

If the universe were a creation within God's mind, it would explain how miracles work. We can work 'miracles' in our mind but they wouldn't be real. Unlike hypnotism in human experience, God's creation is real, objective not subjective. Everybody sees the same reality. To explain further, the universe may not have an independent existence; it may exist not only within God's mind but also within our minds. We are locked into the system. Nature is perfectly logical, but in four known, and probably more dimensions. If we could control our minds without collateral damage, raising our level of concentration, we could master any skill, at least to a higher level.

More about science

The farthest we can possibly see is about ten billion light years and the view would be about ten billion years old (not 13 or 14, or 28 as recent red shifts from distant quasars indicate). The effect of the curvature of space-time on electro-magnetic wavelength increases it exponentially to infinity at about that distance from any observer making objects invisible. The writer is making a rough estimation based on comparison of calculations of distance (red shift and brightness) from observations of Supernova 1997ff described later.

What this is meant to convey is that the lengthening of wavelengths of light, including all electromagnetic wavelengths of which light forms a small part, emanating from the cosmic horizon is making them invisible or undetectable. X-ray emitting bodies, such as newly formed (or forming) galaxies, would be among the last objects to be visible before nothing could be seen. This would be the observer's cosmic horizon. If shorter wavelengths were emitted they would come into visible or detectable range for a time, and then they would rapidly disappear because the expansion of wavelength at that distance would be extremely exponential. Extremely long background (low temperature) radiation possibly comes from the theoretical limit of the cosmic horizon.

Quantum world

Scientists doubt if it is possible to logically predict events within the quantum world of the atom, but we don't know enough (and don't seem able to find out enough) to understand the complicated processes. Nature's rules governing quantum space and phenomena seem disconnected from the larger world, as in communication or movement of electrons occurring instantaneously. The writer's view is that nothing happens, or can happen, without a cause. That is the basic rule of nature. As Einstein said, "God does not play dice with the universe". This was Einstein's response to probability theory about quantum events. It might be impossible for us to predict quantum events but that doesn't mean that they are not logical and predictable to someone with superior knowledge.

Chapter 13

Broad picture reveals the truth Whole duty of man Hitler Holy Trinity Art

Nature's orderliness

Scientists have found that nature's laws work perfectly. God's works are perfect. According to the laws of mathematics, no calculation or equation can produce a wrong conclusion at any stage unless there is an error within the system. Every physical interaction is like an equation in mathematics.

Everything that happens in the universe is 'right', which means there is no chaos. Natural disasters are not accidents to the divine mind. There are no 'accidents' or strange coincidences. There are no hypothetical situations. Everything is justified. If it wasn't God could work a miracle and make it right. Even divine intervention is part of the divine plan. Since God is perfect (by definition) his miracles would also be justified.

Look at the broad picture to get a better view of reality

It helps to understand this surprising claim better if we look at nature (or 'fate') from a broad perspective rather than individual incidents. Thousands of incidents make up a broad perspective. Dwelling on separate incidents is like not seeing the wood for the trees. We do not have the knowledge to see the justification for the trauma of an earthquake or plague, and the more detail we look at the more confusing the picture. But in the broad or general scheme the justification is easier to perceive or understand. God's point of view is not the same as an individual's point of view.

Even so, the writer believes that perfect justice is also done in the minutest detail. Christian behavior is for the children of God the Father. God, because of his omniscience and omnipotence has a different agenda, meaning he has universal responsibility for maintaining perfect logic or reason without which chaos (or worse) would rule.

Another 'broad view' is often found in life experiences. Sometimes we seem to be lucky or blessed, and sometimes unlucky. These events should be seen in the broad view. A series of bad luck events may be followed by very good luck from an entirely different source. Therefore, be philosophical (not angry) about the bad luck when followed by good luck. It may be that God has worked a minor (unprovable) miracle for you. If you are no worse off, be grateful.

It is impossible to distinguish between a thought creation that is perfectly logical and a concrete creation that is also perfectly logical. This presupposes that logic, applied to creation design, can be extrapolated indefinitely to the level where the two types of design are indistinguishable. Our current knowledge of science, especially physics with extra dimensions, including little understood quantum physics, indicates that anything is possible, especially a Supreme Being as head of the body of the universe, without whom the system probably could not function efficiently and not produce sentience. The Supreme Being, and the creation of all kinds of life, would also give a sense of purpose to the whole of creation.

Why does God hide himself from us?

To create a universe without rules of justification (logic) would be irresponsible. Probably, a thought creation, as opposed to a concrete creation, is necessary to allow the development of thinking beings with a sense of self and the intuitive sense of an immanent Supreme Being. It might seem unreasonable for God to make himself invisible to much of mankind, or make belief in his existence unsure, but it is a subject of importance. There must be important reasons for this policy. One would be to test man's perception and faith. Historically, man has always felt intuitively that some kind of spirit or god is watching and guiding our destiny. The absence of the knowledge of God gives man the sense of freedom to express his self-will that he yearns, but it is not without its penalties if he chooses unwisely. Perhaps it is to test man's true feelings. We should be good by nature rather than because of fear, and we should make the effort to think for ourselves and then compare it to the perceived will of God. Essentially, we should show initiative when it feels right, but think how God, or anyone else, is likely to feel about it.

Awareness in a physical universe God identifies himself with nature

God can create a 'mind universe' with rules the same as a physical universe but with a spiritual element that allows miracles. But a totally non-spiritual physical universe would not possess a mind to decide when to override the physical rules. It is doubtful that 'awareness' could exist in a purely physical universe. The universe could not create itself (as it would not exist in the time before creation) therefore it never had a beginning. It has existed for an infinity of time. The fact that it still exists means that it regenerates and is immortal. God's mind would be perfectly logical like nature, but God would always be justified (without sin or error). Sin would be a moral error. To be logical is to reason without error. There can be no effect without a cause. Man is not able to logically define a finite age of the universe. The universe is physical (or so it appears to us). God is mind and /or spirit. God identifies himself with nature, using the pronoun 'I' when referring to nature's works.

If we were to live a godly life without sin we would, in effect, become a spiritual being living in God's reality. If we were to remain on earth we would have heavenly powers like Jesus. We could never exceed Jesus in divinity because he was, is and always has been God's Son. He said, "Before Abraham was, I am". God said to Moses, "I AM". Part of the Christian doctrine is that we become the spiritual body of Jesus Christ in the sacrament of Holy Communion. Thus, heaven and earth become one in the view of the perfect human being. To remind ourselves from time to time, the writer suggests that we 'walk with God in the spirit or body of Jesus' and 'concentrate on what we should be doing'. The second rule is of the first importance. It is unfortunate that such reminders seem to disappear just when we most need them, but it helps if we practice them when we can. 'Practice makes perfect'. We should know intuitively considering our abilities and responsibilities what we 'should be doing'.

Jesus the Word of the Father

When Jesus has appeared, after the first forty days, to various people (eg St Paul) in visions he doesn't appear in bodily form but as a bright light. The idea that we can be 'the spiritual body of Jesus' is a difficult one to imagine. God and Jesus are 'one' in the sense that Jesus is the 'present finite expression' or 'word' of the infinite Father. God is the 'head' of Jesus. Jesus takes his personality from the guidance of God. Jesus surrenders his self-will to God who is omnipotent, omniscient and omnipresent. Christians have the same relationship to Jesus, so Jesus becomes the head of man. But man is imperfect (sinful) so there has to be a compromise. Man lives his allotted span trying to be the perfect Christian then dies and his 'spirit' is taken into heaven from whence he will be transformed into a god-like being on earth in a future kingdom of God.

The whole duty of man is not what some people think

There is one very important point to make in this explanation. The whole duty of man is not to emulate God, but Jesus. God's work is to rule the universe, our work is to please God by obedience to his will, which Jesus has shown by example and teaching that he received from the Father. The difference this makes is considerable. It requires faith in God that he will justify all things. We are not intended to do what would be evil things to us, but let God in his wisdom do what is required his way, not ours. This requires self-restraint especially when we are angry. Ancient kings and rulers often felt like God and demanded servitude from everybody within their power, but Jesus taught us to serve, not dominate and oppress, and so on. Other religions try to serve God but they don't agree with the teaching of Jesus.

In the predatory animal kingdom the tendency is to seize power and establish a pecking order, with the strong at the top and the weak at the bottom. In the more primitive human societies a similar order seems natural, with warlords or military dictators and chiefs (with varying degrees of compassion) at the top exploiting the lower or weaker orders. But it isn't Christian. A higher order of man seeks a better society. The message of Jesus is to put this into practice according to his teaching and example and trust in God to deal with the 'world' in his own infinitely wise and just way. With the power of resurrection God's justice will be done – by God, not man.

If we have Christian love we have empathy. If we have neither we are spiritually dead. Empathy prevents us from doing cruel things, or neglecting our duty to others. Some people seem to be born without the gift of empathy. Empathy, identifying ourselves with another person's feelings forms our conscience. Without conscience we are dangerous creatures, not human in the best sense of the word. If we suppress our empathy for what appears to be a higher purpose, it is probably wrong and ungodly.

Hitler

A few words about Hitler who virtually single-handedly caused the Second World War: He acted as though he were God, deciding who should live and who should die. He was probably doing what he thought was right with (or without) God's blessing. France and Britain were much to blame for allowing him to create a war machine that soon overcame Europe. He showed poor judgment by invading the most formidable country, in size and climate, on earth, with an insufficiently equipped and inadequate army. Had he been a little more patient Germany would probably have been the first country to possess a nuclear arsenal.

Several scientists with first hand knowledge, including Einstein, warned America about the potential of nuclear weapons, so America became the first instead. It is evident that another Hitler could arise with similar ambitions. Political, ethnic, cultural and religious consciousness, pride, and anger as well as aggressive nationalism in a competitive world can motivate people to go to war. One doesn't hear of people blaming mostly themselves for their troubles, but what can a few powerless individuals without leadership qualities do to change things?

Moral values and worldliness

There are two kingdoms in which we simultaneously live: the physical world with its 'worldliness', and the spiritual or moral world with its conscience and empathy with other living beings. We impose our moral values through our world-view to a degree, except when we are tempted and lose the control of our 'higher' emotions and we become subject to our lower or primitive emotions and become, as the Bible says, 'slaves of sin'.

Then there is the eternal question: is there a God? And if so, to what religion do we turn, if any? And what code of behavior are we exhorted to adopt on pain of death and hell or reward in heaven? The writer discusses these questions from a scientific and logical viewpoint.

The Holy Trinity

God the Father in the Holy Trinity is analogous to the subconscious part of the brain in man. The self-conscious ego in man, the part we call 'me', is only a small part of what we really are. The subconscious is the real driving force in the personality. It precedes the self-conscious part in thought and memory and works nearly instantaneously. It converts the information it receives, such as hidden correlations that the conscious can't recognize, into emotion or sense of beauty or other appropriate emotion for the conscious to experience. We cannot consciously create emotion, it has to come to us. Objectivity (observing without being emotionally involved) decreases emotion while subjectivity (absorption) increases it.

The reason there is an ego, that is a single point of view such as we experience, is because it is incompatible with a multiple point of view, and multitasking memory or library of knowledge which describes the subconscious. The ego, as a living being, needs to respond to its ever-changing local challenges. The knowledge it requires to cope with the changing environment is mostly supplied by the subconscious. The words we speak come to us from the subconscious instantaneously, and most of our memories plus many other functions upon which we rely. The capacity of the subconscious exceeds the conscious in complexity (recognition of correlations or patterns – evidence of intelligence) and speed.

The divine mind contains the knowledge of the whole universe because it fills the whole universe. The divine 'Son', as an incarnate being with an individual point of view, would not possess or need 'all knowledge', but would possess all *necessary* knowledge to respond correctly (in harmony with God's will) to all present situations. God the Father (the mind behind reality) anyway controls all situations (the environment). God is the only one that has the knowledge and capacity to understand everything. Thus, without God, man must always be at a gross disadvantage in deciding the right course to take.

The key words that define the difference between the Son and the Father in Christian doctrine are 'infinite' and 'finite'. Thus the Son is the finite local presence of the infinite Father. Since all is in control of the Father there can be any number of finite presences or incarnations of the Son. The Son would have the power of God, but only to do the Father's will. The Son would neither do nor want to do anything that was not the will of the Father, for that would be a sin.

The self-conscious is the 'son' of the subconscious analogous to the Son in the Holy Trinity. The nervous system (from which, in man, the whole brain is made) is analogous to the Holy Spirit, the communicator and executor of divine will. "But the Counselor, the Holy Spirit, whom the Father will send in my name, will teach you all things and will remind you of everything I have said to you." (John 14. 26). "Anyone who speaks a word against the Son of Man will be forgiven, but anyone who speaks against the Holy Spirit will not be forgiven, either in this age or in the age to come." (Matthew 12. 32). " When Paul placed his hands on them, the Holy Spirit came on them, and they spoke in other languages and prophesied." (Acts 19:6).

This is why God said that he made man in his image. We are in his image in principle. If we think of the universe as the body of God that he manipulates as we manipulate our own body, that is another similarity in principle, though we can't perform miracles unless we become divine, which really means 'oneness' or communion in the spirit with God.

To show that Jesus does not compete with God as some people suggest by saying that Christians worship three Gods: He prayed to God, spoke of God in the second or third person, praised God, received his teaching from God, had perfect faith in God, ascended to God from the Cross, preached about God, said he received power from God and surrendered his life and will to God.

Logic, art and the subconscious

'Logic' here means much more than the simple meaning usually attributed to it because it can be extrapolated or extended to describe complex patterns or correlations that we experience as art, harmony (and disharmony), beauty or emotion in all its forms. Beauty lies not only in the eye of the beholder, virtuous or corrupt though the eye may be, but also in its real (not wrongly imagined) compilation or construction as seen by the perfect eye of God. We cannot always analyze what makes a scene or situation beautiful or meaningful but the subconscious usually can.

It might seem a novel idea, that appreciation of art or the feeling of emotion can be explained in terms of logic, but many books have been written on the subject, explaining or trying to explain, why a work of art is to be admired. It is the meaningful correlation between parts that make a work of art or an emotional scene etc. have an effect on the conscious mind.

A simple example can be found in music where harmony can be explained in mathematical terms based on the number of vibrations of a note per unit of time and how this reinforces or clashes with the rate of vibrations of another note. The feeling of pleasure, or displeasure, we experience in the conscious (ego) comes from the subconscious part of the brain. The subconscious converts recognition of correlations or patterns into emotion for the conscious ego. Perhaps the simplest demonstration of correlation is the drumbeat in which the time intervals are all the same length, or form the same pattern. There is a meaningful correlation between parts. I expect that we all wish our subconscious would work in harmony with our conscious ego a lot better than it usually does.

There is no way of proving if the above descriptions of how God works is correct, but the principle that God is in complete control is correct. We could think of the universe as the brain or body of God. It does not matter to us how God accomplishes his will, only that he does. Believing in this (God's omnipotence, omnipresence and omniscience) alters our view of reality or 'the world' (worldliness) and how we react to it. The rightness, or justification, of Christian ethic depends on it. With this knowledge, we need to know how to properly respond to the all-important God.

This is the field of religion. If we don't respond correctly, then suffering and death follow as a logical consequence because we have sinned. If we do respond correctly, then we eventually become 'god-men' 'as angels' or 'like angels' in the words of Jesus (Mt 22. 28, 29) and live a heavenly life on (a possibly new) earth. 'On earth' is the writer's conclusion, as is contemplation that the future body would have physical characteristics such as of the mythical Greek gods. The writer has had one experience (a long time ago) of the remarkable power of the mind to effect dramatic changes in the body (his own), but the experiment proved nearly lethal with long after-effects; there are no shortcuts to self-improvement.

If the universe appears godless or indifferent to the scientific mind, remember that God is omniscient, perfectly just (logical like nature), and omnipotent. He is responsible for the future and maintaining an everlasting dynamic universe. He can make apparent wrongs right. The human mind is only concerned with present time and place. It is also notoriously incompetent. Wisdom and divinity will come to humanity at the appropriate time, but only after a period of trial. All people are sinful (error prone and self-willed) before God and their judgment is likewise imperfect. We need to be aware of this and be open-minded.

Sinfulness and loss of enlightenment

There is a connection between sinfulness and lack of enlightenment. Sinfulness prevents us achieving our full, or divine, potential. That is why we cannot receive the full enlightenment of Jesus Christ, why we are human and not divine. We are handicapped from birth because of our genetic heritage. It is our duty, and in our self-interest, to do our best to live without sin. Sin is ungodliness as defined by Jesus in the New Testament. There are things we should do, and things we shouldn't. It is said that further enlightenment comes when we are ready to receive it. We are all at different stages of enlightenment. Spiritual advancement and seeking perfection in thought, word and deed in harmony with life around us is an unending discipline.

It is part of church doctrine to speak of Jesus taking his manhood into the Godhead. Thus he became divine with divine knowledge and power in communion with God the Father. The same principle applies to man, but we are all contaminated by sin. The sense of self may also change. The memories would remain but in reality we would become a new person, with a different relationship to everyone, including former acquaintances or family. Our code of behavior would align with God, to do otherwise would be wrong, in other words a sin.

Elijah taken up into heaven

To illustrate the 'migration' or transmission of personality consider the following concerning the great prophet Elijah who did not die but was taken into heaven in a chariot of fire: Malachi 4. 5: "See, I will send you the prophet Elijah before that great and dreadful day of the Lord" (God speaking). Mt 11. 14: Jesus speaking concerning John the Baptist: "If you will receive it he is Elijah who is to come". As we change throughout our lives, what determines who or what we truly are? The Jews are not convinced that Jesus is the prophesied Messiah, but if they wait until the Second Coming it will be too late.

No reward without effort

It seems that there is a rule of divine logic that there is no reward without effort or sacrifice. People ask, "If God is good, why doesn't he remove evil and suffering from the world and make us all happy now"? But that is not the way God works. He does it the hard way and the perfect way, by his standards, not ours. Creation is designed by God to achieve a desired result: heaven and earth eventually populated by heavenly beings, born, not created. Then there is the rest of creation, all subject to the same laws and governance, all living, changing and taking their part in their own worlds.

Intelligence

Do not be deceived by so-called intelligence. A high IQ (intelligence quotient) is no guarantee of good judgment. In fact, a person with a combination of high intelligence and bad judgment is recognized by people, educated to assess these things (personnel selectors) as a person to avoid. It is a good thing to be bright as long as they use their gift wisely.

Independence and the teaching of Jesus

If God does not make things plain and reveal his existence there are good reasons for it. One reason is that God wants us to have faith. Another is that he wants us to feel independent and responsible. He wants us to make our own decisions and reap what we sow. The purpose of creation is drama and interest:

life with all its joys and sorrows, trials, ambitions, problems, temptations and success's. In the end he will reveal himself by divine communion to the survivors, the elect or 'saved'. Then humankind will join the rest of divinely enlightened life in the universe as divine beings. All will dwell in heavenly harmony, and all will have the capacity or talent to fulfil their part in God's plan as it unfolds. Jesus should be our guide. Our thoughts and deeds should be linked to Jesus as a reference standard.

The teaching of Jesus on earth is also the teaching of heaven where all live according to Christian precepts. If they didn't, it wouldn't be heavenly. Whoever teaches against Jesus is in error. Life on earth among bad people reflects hell rather than heaven. Societies make their own hell, and the rulers who perpetuate the conditions are the personification of the devil.

The teaching of Jesus sounds one sided to the human ear: there is no coercion, punishment or death threats against evil doers, at least, not in this life but certainly in the next. But the message is to Christian followers, not governments. God reserves the right to punish and kill whether by natural means or man. God works in mysterious ways to man's mind. God controls the environment and everything that happens. Christians have nothing to fear, especially death.

Einstein has never been proved wrong.

The theory of relativity is confirmed as most scientists admit. Einstein has never been proved wrong. The idea that science is incompatible with religious belief is because people do not understand divine revelation or contemporary physics. Divine revelation is not black and white, always right in the scientific sense. It is appropriate to the people of the region at the time, in other words 'relative to the observer' which means it is only as 'good' as the observers or prophets. We have to look carefully at prophets (the 'observers') and the development of the region and its people. Scientists say they believe Einstein, but only in demonstrable experiments. In universal space matters they are only paying lip service because they haven't understood that part of his theories. Einstein didn't understand that part himself but he didn't have the advantage of modern discoveries.

Culture and religion

Culture is related to character, and character is related to religion. Is it not an historical fact that religions are related to cultures? What does this make of God? If God cannot sin or err then it is right that cruel and insensitive people are drawn to harsh religions, and humanitarian people are drawn to humanitarian enlightened religions. Note that God does, in his own way, what man is not equipped to do, because God is omniscient and omnipotent which man is not. Only God is the perfect judge. "Behold, I have created the smith who blows the fire of coals, and produces a weapon for its purpose. I have also created the ravager to destroy" (Isaiah 54. 16). God identifies himself with what we would call nature.

God does not do evil things, but he allows evil people to do evil things, then allows nature to follow with natural consequences, suffering and death. In this world of sin death must come that others may live. Animals would rapidly breed and plants would spread thousands of seeds until there was no room left or food for survival were it not for death. God has sent a Savior so that the elect may become immortal in heaven – or an earth that is heavenly, the Kingdom of God.

Different regions with different climates have an effect on human culture. The first essential for life is water, then soil, light and warmth. Temperate regions tend to produce planners who store food and fuel for the winter. The equatorial regions with heat and water and virtually no winter burst with life and tend to produce a human population who have no need to plan for the future but have to cope with strong competition. The parts in between, with deserts and varied terrain have a variety of problems. Mankind's main long-term problem is competition with people. It is believed that some regions, such as the Middle East used to have more coverage of forest, which cooled the area and brought more rain. Science will eventually devise a way to control earth's climate.

Suggestions of corruption of biblical records

It has been suggested that the New Testament account of Jesus has been added-to over time to impress readers, but the idea is untenable. Who could possibly invent the idea of a man being chosen as earth's Savior? And give sermons in the style of Christianity? 'Love your enemies' would have been laughed at if it hadn't been Jesus preaching. No one had any concept of the Holy Trinity until Jesus arrived. Without his miracles no one would have listened to him. His final miracle, his resurrection after crucifixion, would have been unbelievable if it hadn't really happened. His followers were witnesses, and they acted appropriately. No one could have spoken consistently in the same manner and appear after his resurrection to many. He revealed himself to Saul, an ardent Jewish persecutor of

Christians, who changed his name to Paul and became the greatest contributor to the New Testament of the Bible. Also, any minor inconsistencies or unflattering events would have been changed. No Holy Scripture has been more subjected to examination by thousands of scholars over a very long period than the Bible.

Christianity has become the foundation of the highest Western standards of civilization: consideration for others, good manners, generosity, highest ethical standards, purity, forgiveness, selfless love for our neighbour, longsuffering, understanding and so on in the same vein. Above all else, hope of salvation gives us motivation to be Christian in deed as well as word. Our justification is faith: faith in Jesus as well as God who sent Jesus.

From a political point of view, regarding hostility between leaders of nations, lack of Christian principles causes lack of cooperation. A Christian society is, above all else, supportive and cooperative; otherwise we become like the house of evil, divided amongst ourselves. If we haven't got the sense to be cooperative for our mutual good, willing to compromise, and we are armed with thousands of weapons of mass destruction, we could be a disaster waiting to happen. If compromise is regarded as weakness without very good reason it is an unchristian judgment. Jesus is regarded as the Prince of Peace, "blessed are the meek for they shall inherit the earth" (from the Sermon on the Mount, Mt 5. 5).

Motivation to sin

Sin is natural, more so to some people than others. Animal type appetites account for much motivation to sin. Money and power offer a way to satisfy our desires whether they are socially constructive or destructive. Whether we think we were 'created' as literally described in the Old Testament or not, we are both physically and mentally associated with animals; but this behavior is not compatible with Christian teaching. We have to disassociate with the antisocial and unholy animal characteristics and become more civilized and humanitarian. Otherwise, we increase in sin and folly and have no hope of salvation to a higher life.

Chapter 14

Relativity Defence Good manners Western civilization Doomsday prophecy

When the idea of relativity became popular it was not uncommon for some people to justify wrong ideas, such as there is no such thing as absolute right or wrong, or, for that matter, truth. *Everything* was relative to someone or something else. But in science, the exact description of relativistic transformation is described, so that there is no disharmony in the general structure. Misapplication of relativity theory leads to wrong conclusions. Much that is vulgar is praised and rewarded by corrupt people, but that is what God allows in his infinite wisdom. The end of misapplied effort is destruction.

The application of relativity to religious revelation or experience fits historical reality beautifully; look at all of Jewish history: does it fit the local revelation or the universal? Of course, Jehovah blames the disobedience of the Jewish people for their misfortunes, but so would the Universal God; The writer searches for the truth and believes he has found it.

Jews are noted for their wealth as moneylenders but not admired for it. They are also noted for their pre-eminent position in science being awarded Nobel Prizes out of proportion to their population, medicine (they study harder) and attain leading positions in many professions, so much so that Hitler hated them. Einstein was a Jew but he abandoned Jewish religious beliefs early in life. Jews are unique, but still generally unpopular. If this book helps to persuade even some of them to become Christians it would be widely regarded as a miracle. The writer would be immensely pleased.

Defence

The writer does not wish to criticize non-Christian religions, but to show non-Christian followers how to make corrections to their beliefs and way of life. If conditions are such that Christianity cannot be practised for good reason then it is hoped that some moderation will be allowed to prevent unnecessary violence. Although violence is not part of Christianity, defence may be necessary to save people from harm, especially our own families. Defence is easier than offence. God, everpresent, will show a way.

The consensus in the West is that all nations should be, at least, civilized, allowing people their own beliefs without interference as long as they do not illegally interfere with others. Religions that have political agendas and hold power over non-believers can only expect severe opposition. People without sin cannot be destroyed, but who is without sin?

Defense is not within Christian teaching. "Then they took up stones to cast at him: but Jesus hid himself, and went out of the temple, going through the midst of them, and so passed by" (Jn 8. 59). The writer would call this passive defense. He could have used his divine power to destroy them but that would have caused an unwanted reaction. He could have converted all people to Christian belief by exercising his power but that would not have suited God's plan.

Justifiable defense, in the writer's view, is acceptable, but the whole context must be understood to form a correct opinion. We put dangerous criminals in prison to protect other people, that is a form of defense. In the Old Testament, as in most primitive societies even today, war was a way of life between tribal, political, ethnic and national enemies. This cannot be prevented except by raising both the standards of living, behavior and of thinking of individuals who compose the populations of the conflicting countries and factions. People inevitably cause much of their own discontent and misery. If people can't or won't change and adapt, then things will remain as they are until the natural order decides their fate for them.

Love of neighbor foundation of good taste and manners

The difference between 'good taste' and 'bad taste' and all between is in the inherent goodness of people. For instance, genuine good manners show good taste, bad manners show bad taste. Insensitivity towards people or lack of empathy is also bad manners. The writer cannot imagine the kingdom of God populated by people with bad manners. All of the best social relations are based on the teaching of Jesus. The whole future of mankind, if we take the biblical view, is that God (via Jesus) will take the best and make them better, and leave the rest. The writer refers to genuine manners as based on non-judgmental appropriate loving-kindness, respect and consideration.

Who will be saved? Who can judge?

The writer doubts that anyone could choose who will be saved and who will not. God does not judge by appearances and can transform the most evil-seeming person into a saint if he wills. God can neither sin nor err. In the meantime we should keep faith, believe in the Christ Jesus whom God sent for our salvation, and do our best to live by Christian principles. Humility is a principle virtue, and loving-kindness towards our neighbor is one of the two great commandments, the other is to love God.

Writing of judgment, there are two ways we can be judged. The first way is the obvious way: our good deeds can be weighed against our bad deeds by an omniscient and omnipotent God, which is the historical and simple belief; or we can be judged according to our own judgment. If we judge Jesus to be the Christ, our Savior, who taught the truth, was resurrected and has all power over earth to give life as he had power to raise himself to eternal life, we will be saved. Because Jesus can preach to the dead and has divine judgment and knowledge, he can save whom he will (John 5. 25).

The justification of this conclusion, that we seal our own fate regarding salvation, does not give Jesus the power to save, or us, but God; for it is God who gives us our beliefs based upon the kind of people we are. We cannot follow Jesus except it be God's will. Jesus and God are one. The will of Jesus is the will of God, and God cannot sin or err. We should not think we are good, for "none is good except God" (Luke 18. 19). Therefore let us be wise, not foolish, humble, not proud.

Jesus, in heaven, has God's power and knowledge, and has earned the right to overlook our sins and transform us into divine beings. All is according to God's will. Heaven is not a place made for man, but man made for heaven, so we should concentrate on spiritual and moral qualities above other things. Jesus will save if we are forgivable.

The universe is designed for perfect people to enjoy

The universe creates order out of chaos. It fulfills every desire of the perfect mind. Chaos to us is all part of God's perfect plan. If man were perfect he would know all this, and nature would be an extension of himself. Man, in theory, can rise above 'the world' (as Jesus said of himself) and become divine, one with God, but man needs divine help. We may try to be perfect Christians, but we all fail. We need a Savior. Only Christ-seeking, right-thinking people will be saved. Others are following their own corrupt thoughts.

Elegance of nature

The universe is very complicated, which it has to be for the amazing performance that we can see, plus what we cannot see. Scientists of all disciplines, as much as thoughtful artists, are frequently overawed at nature's wonderful design, in the creation of continents and seas to the miracles of birth and growth. And all the matter in the universe is basically made from only one kind of atom: hydrogen, which is the smallest. Elegance, so cherished by mathematicians and physicists, is only one of nature's many attributes; the most important, but mainly unappreciated attribute, is perfection.

Science has made it appear that a God, or belief in God, is unnecessary. But this is because we have only scratched the surface of the problem of understanding nature. As scientific knowledge increases so will our understanding of God increase. We are presently polluting the atmosphere because of our lack of progress in science. We need a breakthrough, a new direction with light at the end of the tunnel. We need to find the secret of unlocking the power in the atom in a controlled way without dangerous radiation. Science has done much good but there is a downside, because power can be used for good or ill. Every discovery that increases man's power over nature also gives man more power to destroy. A growth in faith and understanding needs to match the growth of scientific discovery.

The benefits of science and the real consequences

It has also made possible the increase in population of the earth to a degree that seems excessive. Science has progressed but human nature has lagged behind. People are living longer but the frailties of old age are still with us. Science has produced more food for the poorest people of the world, but the end result is that the growth in population has increased the number that subsist in poverty with a high death rate due to malnutrition, disease, warfare or starvation.

The problem of persistent poverty, despite the progress of science or efficiency of production is, quoting from a newspaper article on starvation, 'Too many mouths to feed'. For instance, Ethiopa's population has increased in twenty-five years from forty million to eighty million. Britain, a relatively rich country, whose population since 1710 has increased from seven million to sixty-one million, who couldn't feed themselves today, can afford to import a third of their food. China recently effectively imposed a limit to families to have only one child, but this has caused, or is seen to cause, problems with a disproportionately ageing population. If man does not find a remedy for these problems then nature/God will.

Armed with the knowledge of God

To be armed with the knowledge of God strengthens our faith and increases our power to resist temptation. The more we resist temptation the more disciplined and pure our minds become, the better to see the truth and know God. We should not feel defeated if a particular vice (any vice is also a sign of sin – the vice is obvious but the sin is more subtle) seems too strong for us. It is a temporary psychological (affecting judgment) or psychosomatic (with physical effects) victory for the tempter. It is our duty to continue to fight against the temptation, not to be defeatist, and have faith that our resolve will be rewarded and our determination strengthened. We are weak but the strength of God is infinite and he is very patient and loving to those who love Jesus. Remember that faith can move mountains (Mt 17. 20).

This is the motivation of saints, and the reason why Jesus, who was without sin, was the true prophet of God. The body is (or 'we are') the temple of God (Jn 2. 19. 1 Cor 6. 19). Ignorance (a consequence of sin) separates us from God. Man is capable of divinity by the grace of God. This is the basis of Christianity.

Apostles

The New Testament of the Bible gives a very full personal description of Christianity, albeit with the human failings of some of the apostles, notably Peter in his judgmental and lethal dealings with Ananias and his wife Sapphira ("and great fear came upon the whole church, and upon all who beard of these things." Acts 5 1-11), and cutting off the ear of a slave of the high priest, which Jesus healed (Jn 18. 10. Lk 22. 50, 51). The apostles of Jesus were ordinary men who became extraordinary.

Leper colony film

The writer recently saw a film about an island sanctuary, or place of separation, for people with leprosy set in a time before a cure was found. A Christian martyr went to the island to care for the physical and spiritual needs of the people, eventually contracted leprosy and died. The atmosphere of the film was, apart from some uplifting hymns, mainly dark and miserable. The good news is this, and should have been the theme of the film, that God sent Jesus Christ to offer forgiveness of sins, salvation and everlasting divine life to all who believe in him, and give convincing physical and spiritual proof that he spoke the truth.

Great light

But we have to have faith in him. Saints may have died in present misery, but spiritually they were happily assured of resurrection into a blissful future life. He rose from death and appeared to his disciples and others for forty days. The Christian faith would not have overcome opposition and grown if it had not a firm foundation. After that time Jesus has appeared to others but not in his earthly body. He appears as light. An old man spoke to me in a supermarket and told me he had seen Jesus as a bright light. He appeared to St Paul, "Suddenly there shone from heaven a great light round about me" (Acts 9. 3).

Ancient Greeks

Ancient Greece in the time of Pythagoras had some people just as intelligent as any modern man; they even worked out with their primitive instruments that the earth must be spherical, but they didn't know about gravity and earth's orbiting of the sun, and the idea died for lack of support.

Two Testaments

Some parts of the Old Testament are not Christian, but it does foretell the coming of a Messiah, and reveals much knowledge of God. If it had been fully Christian, there would have been no need of a New Testament. It was not God who needed two testaments; it was man's development and a world plan for salvation that called for the need. God embraces everything in the universe, but everything must obey the law of logical cause and effect, in other words: righteousness in ethics and truth in science.

Old Testament

The Old Testament has been compared to the roots of a tree, from which grew the New Testament. Jesus quoted liberally from the Old Testament, especially those parts that prophesied his coming. The first verses in Genesis, discussed elsewhere, are demonstrably picturesque or poetical but fundamentally true to an understanding reader. Revealing the truth about Genesis makes the Old Testament more believable, but asserting its literal interpretation makes it less believable. God creates everything but not literally in the way described in Genesis. Even today scientists are discovering things we didn't know about evolution, and the theory needs some refinements.

A person's experience of nature or how the world treats him varies with his personal development and circumstances. Nature is as variable as any person. If the person changes his outlook how the world (nature, and other people) treats him also changes. There is no bad luck or good luck. The system is logical but we probably cannot see it. Even our bodies can miraculously change but this rarely happens. We have limited power to change ourselves, but if we had enough knowledge we could perform wonders. God's ways are not man's ways. Man's ways are simple, quick, and potentially full of error, while the 'mills of God grind slow, but they grind exceeding small'.

One of the most important things a Christian (or anyone) should strive for is justification or 'rightness'. It is a natural law that keeps the universe perfect. If there is a right way, and several wrong ways of doing anything, we should choose the right way. It may not seem to have anything to do with religion but it should be part of the code of conduct of every Christian. It would come under the heading, 'Love God'.

The nature of creation, arising from lower stages of life by whatever the method nature/God employs, guarantees eventual self-destruction, as seen in the present political situation with the creation of nuclear and (probably) biological weapons. In other words, God has put sophisticated weapons in the hands of violent people unfit to possess them. This is not what God would want except to fulfil man's destiny.

Doomsday prophesy

Jesus in the New Testament describes the last days of life on earth as we know it: "For then there will be great tribulation, such has not been since the beginning of the world until now, no, and never will be And if those days had not been shortened, no human being would be saved; but for the sake of the elect those days will be shortened (Mt.24. 21, 22). Thus, sinners will self-destruct, or otherwise be destroyed, while God/Jesus saves the selected ones (the elect) to be raised to a new life on a new earth. The time of the 'last days' appears to draw close, but "no one but the Father knows the time."

The only natural cause that is at present a known threat to the earth is pollution, causing climate change with unknown consequences. If the consequences are severe enough they could cause wars that could escalate.

Why should evil thrive?

In conjunction with the idea of relativity in religion we have to consider why God allows evil to appear to thrive. One suggestion is that God gives the serious sinner enough rope to hang himself. Another is that evil is mutually destructive, nation is pitted against nation which in these days could destroy all life on the planet. Man has freedom of choice in matters of belief or opinion, according to his own judgement, experience and intuitive feelings, apart from uneducated people who have been indoctrinated from birth. It also puts every Christian in a world that is designed to try his faith.

Some evil people do not seem to suffer like many good people. They seem to live a charmed life. Perhaps it is this that gives them the assurance that they are free to do as they please without regard for consequences. If evil people suffered they might turn from their evil ways and grow wise and be saved, but in their hearts they would still be evil. Therefore God saves them from present suffering. The saints suffer and are saved. If we suffer there will be a good reason for it. God punishes those he loves for correction. People turn to God when they suffer.

We also have to remember that the sinner is in bondage to his sin. This is the Achilles' heel of sin; the sinner, in the end, loses control and brings destruction on himself. Christian doctrine would lead to the conclusion that the wilful sinner lacks the humility to heed God that would save him. A preferred belief in a God with an unchristian doctrine would be a natural consequence of an unchristian character. The elect, or 'chosen', respond to truth when they hear it. The sheep know their shepherd. Everything is connected and has a purpose. God is everywhere but no one has seen God. It is wrong (or childish) to think of God in the shape of a man; is he tall or short etc? God is spirit and he is omnipresent.

Chapter 15

Science reveals the greatness of God Entropy Supernova 1977ff Steady state Cable-laying space-ship UFO's

Science, properly understood, reveals the greatness of God. But firstly, we have to understand some of the mystery of nature in order to appreciate the greatness of the divine mind behind nature. The popular scientific view is that entropy must always increase in the universe. No engine can be built that is 100% efficient. Friction creates useless (unavailable) heat. If this theory were to be correct the universe would end its activity in its 'heat death' where everything is of a uniform temperature, and there is no potential left for a flow of energy. But there is no sign of the universe 'running down', and there is no friction in space. If the universe runs like an engine, it is a perfect engine.

Expanding universe theory inadequate

Modern technology keeps uncovering facts that do not fit the expanding universe theory. For instance, the ages of some galaxies exceed the time available since the supposed 'Big

Bang' commencement, and the rate of expansion appears to be increasing with distance rather than decreasing with the constant pull of gravity towards an assumed center. "Explaining why the expansion of the Universe is currently accelerating is certainly the most fascinating question in modern cosmology," says Luigi Guzzo, lead author of a paper in this week's issue of Nature, in which the new results ("Tech Astronomy.com)..March 6 2008. Search 'expanding universe'). Alternative explanations for the red shift are being assiduously researched.

The phrase, 'currently accelerating' really means that the more distant stars are apparently receding faster than they should be according to the dynamics of a 3-dimensional explosion. They are not moving faster than they were twenty years ago. If the writer is correct, they are hardly moving at all. The ubiquitous red shift is due to distance and the curvature of space-time, not motion. All galaxies are moving relatively to each other but not at spectacular speeds.

Steady state

Also, if the universe is spherical as Einstein theorized, gravity from galaxies near the cosmic horizon, would work on the far side of the sphere (along geodesic lines) and eventually pull in the opposite direction to the gravity on the near side. Four-dimensional geometry has some similarities to three. Whatever the outcome of mathematical study gravity has to be diminished exponentially to prevent all the matter in the universe forming one huge ball and eventual collapse. Fortunately, the universe appears to be in a permanently steady state mode.

"A Nasa spacecraft has detected dozens of new-born galaxies in Earth's part of the universe. These unexpected cosmic infants were discovered with the Galaxy Evolution Explorer, which managed to spy them because of the huge amounts of ultraviolet light they emit as they furiously form stars out of gas, astronomers say." (New Zealand Herald, Dec. 23, 2004).

Supernova 1997ff

The universe is spatially a little smaller than the red shift effect indicates. We should soon have a more accurate figure based on other known methods of measurement. The red shift of supernova 1997ff (Internet search: supernova 1997ff) indicates that it is 10 billion light years distant but its known brightness (from observation of similar supernovas) indicates 7 billion, which is another confirmation of the relativity theory.

There is now evidence that we can see almost the entire universe and we appear to be at the center, which would be extremely unlikely in 3-dimensional space, but not in a 4-dimensional continuum, which has no common center, as Einstein's theory shows. There are several other objections that prominent cosmologists raise from time to time, but no one seems to suggest a credible alternative explanation. Every observer sees himself at the center of the universe, which we do. Even with the Hubble telescope (2.5m) and certainly with the much larger (6.5m) James Webb space telescope now in preparation for launch in 2013, we can see the cosmic horizon. We just haven't recognized it, probably because astronomers do not know what to look for.

The shape of the universe

The writer's special interest is in physics and cosmology with emphasis on Einstein's theory of 4-dimensional space-time. We cannot straighten or shorten the line we travel in space any more that we can shorten the line we travel on earth, without boring a very long and deep tunnel. On earth we would only make the journey longer or shoot off into space. In space we would have to travel in time, as in a time machine by travelling very fast. We cannot travel to the past, but we can travel to the future, only to find it has become the destination's present.

Cable laying spaceship

To illustrate the nature of the shape of the universe, imagine a cable-laying spaceship travelling at light speed. It travels in a subjectively straight line, which is really curved in both space and time, around the universe. Arriving at the same space it left, the end of the cable would meet the beginning forming a great circle 20 billion light years long, but the end of the cable would be new whereas the beginning of the cable would be 20 billion years old. This is because, at light speed, the ship and its contents don't experience the passage of time because of the contraction of the time frame to zero.

The velocity of light

Just to show the apparently illogical nature of our 4 dimensions, the observed speed of light is the same however and wherever we measure it. Motion in any direction of the light source or measuring equipment makes no difference to its observed velocity. Light, in its own time frame, travels instantaneously (at infinite speed). Light exists in a constant state of instantaneous motion in our three spatial dimensions. Its *observed* finite speed (always a constant 'c': 300,000 km per sec.) is governed, or caused, by the space-time curvature of the universe, but the writer doubts that anyone else knows this so readers should not expect confirmation.

Light always moves at the same speed to any observer. This is true of light in a vacuum. When light moves through glass or water it appears to slow down but this is due to molecular action in the material transmitting the light. In the space between atoms and molecules light still travels at 'c'. In its own time frame it travels at infinite speed because it exists in a timeless dimension (zero time frame).

Einstein proved mathematically that when an object accelerates towards (observed) light-speed it increases in mass exponentially towards infinity, shortens in the direction of motion, and its clocks (physical processes) or time rate slow down towards zero. Matter can never attain light speed as already explained. Acceleration implies a change in speed over time. Instantaneous motion means no time for acceleration.

It follows that light exists in a timeless dimension because its time rate has contracted to zero, and it moves instantaneously, but only in its own time frame. Space is integrated with the time dimension forming a space-time continuum with some weird geometrical properties that only become noticeable at extreme speeds and distances. Light has no 'rest mass' (mass of a body that is not moving relative to its observer) but it has mass by virtue of its energy (Feynman). Energy is another state of matter with unique and very different properties that appears to co-exist in a timeless dimension, yet it is an alternative form of matter.

Distance changes time

Distance means a change in time as well as space. Clocks in different places show different times Light from the sun takes 8 minutes to arrive here which makes us think that light has a finite speed, but the delay is due to the construction of space-time. The sun really is 8 minutes behind our present time, just as we are 8 minutes behind the sun's present time. Light connects us instantaneously in its own time frame or system, but it is only showing the truth.

Time travel

Time is infinite, the future is constantly being created and the past is creating the present. We could travel to the future (by accelerating to a very high speed) but we cannot travel to the past. Each observer is at the spearhead (present) of the creation of the future; everywhere else is in his past. The potential infinity of time means that the universe is, or can be, everlasting. The indestructibility of mass/energy means that it must be everlasting.

Universe constantly dying – or like the legendary Phoenix

The universe is constantly dying but, like the legendary Phoenix, arises from the 'ashes', actually electromagnetic energy, reforming into new matter. Most things occur very slowly in our time-scale. The total mass of matter in the universe is probably fairly constant. Galaxies are constantly being created, but we cannot see them until they become hot (bright) enough, as the huge clouds of gas swirl in decreasing circles finally concentrating into new galaxies. No two galaxies, suns or planets are the same. Nature/God delights in variety.

FTL travel

It is widely believed that nothing can exceed the speed of light. However, a space ship can exceed the *observed* speed of light without limit but only in its own time frame. Observers would never see it attaining light speed. If the ship accelerated enough, and one hour in an earth observer's time equalled one minute in the ship's time, then the earth observer would only see the ship travelling at one sixtieth of its (subjective to itself) true speed. The ship would see earth spinning, and the moon revolving, sixty times faster. FTL = faster than light.

This is in accordance with Einstein's theory (time contracts with acceleration). If we could see the passengers inside the ship they would appear to be moving sixty times slower. The theory could be tested with an accelerated camera or specialized equipment, observing a binary star for instance, but nobody seems to have even tried it. By these criteria, the ship must exceed the speed of light in the view of its passengers, but in our (observer's) view it doesn't. The ship, at any sub-light speed, would see light moving at its normal constant relative to themselves.

If a father went on a spaceship and travelled at very near observed light speed and arrived back after 20 universal years, his son would have aged 20 years while the father had hardly aged at all. The father would have only experienced a few months of time, depending on his speed (which can only be gained with acceleration).

Incidentally, I don't think that we can travel much in space until we find a force that acts like gravity on the whole body, as in 'free-fall'. There is no sensation of acceleration if there is no inertial resistance. With such a force there would be no bodily limit to rate of acceleration or deceleration. If reports of UFO's are true, such a force exists. The reality of UFO's is regarded as highly controversial. One sighting from an airfield of a UFO that nearly landed and then took off (and was observed on radar) reported a speed of 3,500 miles per hour.

Life on planets

NASA officials and other leading scientists at the annual American Astronomical Society conference agreed that planets like earth where life could develop probably are plentiful and should be discovered within five years (News Jan 9 2010). Their evolutionary history would probably run parallel to what ours is prophesied to be: stormy, disastrous followed by salvation of a few chosen beings. The writer doubts that anyonc has seen the passengers in UFO's.

Foreshortening of space

There is also the contraction in the direction of motion of the accelerated body or ship. As a ship accelerates the passengers do not experience any relativistic transformation within the ship, but their observation of the outside universe does change. The universe appears to be foreshortened.

The foreshortening effect applies to both observers on ship and earth, because motion (the result of acceleration) is only relative to other objects and can be regarded as reversible (either one could be the one moving, as taught in 'University Physics' 9th edition, Young-Freedman, p 1205). Both observers see each other as foreshortened. The ship should see the distance to their destination decreasing (shortening) as they accelerate. The ship itself would shorten but the passengers would not perceive that, because everything in the ship is in the same space-time frame.

It appears that the ship does not have to exceed light speed in its new space-time frame because the distance has contracted instead, as has the rest of the universe, in which everything spherical would appear oval (foreshortened in the direction of motion).

Kinetic energy is energy of motion gained by acceleration and increases as the square of its speed. Kinetic energy converts to force when moving cars collide. Mass and weight are the same at the earth's surface. In deep space the same mass would have no weight because there is no gravity, but it would have the same mass or momentum. A moving mass could do damage to another mass in proportion to the square of its speed and in direct proportion to its mass the same as on earth.

The energy of acceleration must be stored to maintain the mass/energy constant in the universe. We think of increased kinetic energy as energy of motion or momentum, but in 4-dimensional space-time this is not entirely correct. It is practical at terrestrial speeds because the relativistic changes are extremely minute (a small increase in mass stores, or is equivalent to, a very large amount of energy) but increasingly impractical at close to light speeds when the mass-increase factor becomes significant. The scientific community accepts the mass increase as a proven fact, though I doubt that anyone knows the reason for it, for I have never heard or seen it mentioned. It is another confirmation of the 4-dimensional space-time continuum nature of the universe. Kinetic energy increases as the square of the speed. A car that doubled its speed could do four times the damage if it crashed.

Chapter 16

Cosmic rays and Lord Rutherford Lorentz transformations Time and space travel Quantum physics Interstellar flight

Cosmic rays are a constant stream of very high-energy (very fast-moving) particles that bombard the earth from space. The primary radiation consists of nuclei of atoms that collide with atomic nuclei in the earth's atmosphere producing secondary radiation. The surprising discovery of the primary radiation's depth of penetration within the surface of the earth was that it was up to eleven times greater than was expected. However, with Einstein's explanation of mass increase with acceleration (giving the particles more momentum) the mystery was solved. Two students of Lord Rutherford, the New Zealander who discovered and named alpha, beta and gamma rays, who first split an atomic nucleus, used cosmic rays in the same way that modern physicists use huge particle accelerators.

Space travel

How is the space ship's time passing in relation to earth observers? This is the important question because it determines the human limits of space travel in our enormous universe. Let us say that the ship takes 4.1 years in our time to travel 4 light years (the approximate distance to our nearest star). Time contracts on the ship to one fortieth, making the travellers' subjective time for the journey on the ship 0.1 year. We know by theory and experiment that clocks slow down in accelerated objects. I have used this conclusion to explain how galactic travel is possible in reasonable periods of time.

The following is a short table of speeds as percentages of light speed (of a space-ship), and the equivalent relativistic transformation factors, affecting mass (+), length (-) and time (-): 40% = mass+10%; 80% = +67%; 99% = +700%; 99.99% = +7000% (70 times rest mass, 1/70 length, 1/70 time). Rest mass is the mass relative to the observer when there is no difference in motion between them. All three factors change when acceleration takes place, but only to other objects (or observers) whose motion has not changed, or has but at a different rate.

It should be understood that 40% of light speed is 120,000 km per second and only produces 10% increase in mass, 1/10th decrease in length and 1/10th decrease in rate of time. There are 3600 seconds in 1 hour, so 40% of light-speed is 432 million km or nearly 270 million miles per hour. Travel to our nearest star (with possible planets) 4 light-years away at that speed would take 10 years.

If all this dwelling on the vastness of the universe makes man's affairs appear insignificant I will repeat a story from 'The Contemporary Christian' by John R W Stott (Inter University Press 1992 --2002,): One scientist said to another, 'astronomically speaking man is infinitesimally small'. 'That is so,' responded his colleague, 'but then, astronomically speaking, man is the astronomer.' I should also add that nothing is insignificant to God or nature. The smallest event can have an ever-growing influence on the future. I recommend John Stott's book especially for its Study Guide and for its evident clear-sightedness.

Spooky action

There are lots of experiments at the quantum level where light appears to move instantaneously to everyone's surprise and mystification. One is described in the magazine Science (AAAS) March 17, 2000 pp 1909-1920 called 'Spooky Action'. It is possible that this anomaly is caused by the system of timing, because time does not exist to electromagnetic phenomena. The mysteries of quantum mechanics which scientists gave up trying to solve a long time ago in favour of probabilities (random individual results) might yet be understood. In the well-publicized argument about the unpredictable motion of individual electrons, I support Einstein's sentiment that "God does not play dice with the universe". Electrons must follow the line of least resistance; nothing happens without a cause.

Quantum physics

On the subject of 'quantum computers' it was reported on BBC World News that scientists had discovered that a change in one electron produced an instantaneous change in another paired electron a meter away instantaneously. This indicates that a computer could be made to produce instantaneous, or millions of times faster, calculations (Jan 23 2009, 2.15 pm NZ time). Another newspaper report states that scientists are on the brink of a breakthrough in quantum computer technology based on electron spin - spin up or spin down. The word 'spin' appears to be more a convenience title than a conceivable reality, but it serves its purpose.

Quantum physics has been long regarded as nearly a total mystery because it defies known rules of physics and of commonsense. Subatomic particles seem to exist in more than one place at the same time, their position is unpredictable and they affect each other simultaneously - another example of the mysteries of nature. It is possible that measurements at the quantum (extremely small, atomic) level are affected by the construction if the atom. Our mental picture of a positive nucleus with electrons revolving around it is over-simplified. They don't revolve like planets around a sun, instead, to the writer, they have the nature of electrical charges that 'exchange properties' instantaneously rather than move.

It seems that, within the construction of an atom (the quantum world), time has no effect as it does elsewhere. When light moves through space it is reduced (to the observer) to 'c', the constant speed of light. But it is information about a state of change in one part of the atom to another paired part that is instantaneous. What is the nature of this 'information'? What can travel instantaneously in our time frame? So far the answer is nothing, except between electrons in a single atom. Tunnelling, as in transistors, is also instantaneous. It seems to be a field worthy of research. Who knows where it might lead? What the writer does know is that man's knowledge of quantum science must be relatively primitive.

On reflection, transmitting electric current along conducting wire is like sending water along a pipe already full of water. We push water into one end of the pipe (water is virtually incompressible) and water is pushed out of the other end, but it is not the same water. The same with electricity, what comes out is not the same electricity (electrons) as went in. The speed of electricity through a conductor is a little less than the speed of light. In a transistor (a kind of electronic valve that controls flow) some of the current that goes in goes through and does its work. These are presumably the *same* electrons. In the atom, the same principle applies. The same electrons 'move about or around' the core of the atom also instantaneously or they change their attributes instantaneously. This doesn't mean that they are in constant motion.

The Lorentz Transformation Formula

The mathematical (Lorentz) formula for calculating the factor or rate of relativistic changes (commonly referred to as 'Beta') is very simple: Beta = 1 divided by the square root of 1 – v (the *observed* velocity of the object) squared over (divided by) c (light velocity) squared. It is always between 0 and 1 because we can never observe anything moving faster than light. When the observed object is not moving relative to the observer, v is 0 and the answer is 1, which means there is no change. A full treatment of the Lorentz factor can be found in Wikipedia on the Internet.

On the other hand, if v equalled the velocity of light, which is impossible within our present knowledge of physics, the answer would be 0. Any finite number divided by 0 equals infinity so the degree of change would be infinite. This means that the observed mass would be increased to infinity. The same with time: its clocks, and biological processes, would appear to the observer to have stopped. No motion would be visible within the ship, even though we might observe it for a long time travelling at light speed. Passengers within the ship would arrive at their destination instantly, but a million years would have passed in universal time while the ship completed a million light year journey.

Interstellar flight and UFO's

Now this opens up the possibility of interstellar flight. We are presently limited by time because of the slowness of space flight with our level of technology and the human condition. But there are probably planets populated by 'beings' with very advanced knowledge. Since there is no theoretical limit to the subjective speed of space flight we could be visited by 'UFO's' from our nearest star systems, just over four light years away that, to them, might only take a month, or less. To us, the journey would appear to take a little over four years. If they were capable of transformation into a timeless dimension for the journey they would arrive instantly in their own time frame, but four years in ours.

They don't do anything except affect people's opinions. They show that there are other living beings in the universe that are aware of us, but in a way that is not conclusive to cynics. They are positive evidence of very advanced and capable beings. They do not seem to have a purpose except, to the writer, they are letting us know that they could deliver us from a worldwide catastrophe that would otherwise kill us all.

If I may introduce a philosophical concept, ie if we could be copied, molecule for molecule, into an electronic form, sent on our long journey, then recreated into our original form, we would not know any difference. With the same memories, body and brain, how could we (or anyone else) tell? Our bodies change as we sleep, and greatly change over the years as we age, but we still feel the same person, though with diminished powers.

Gravity equivalent to acceleration

According to Einstein's theory of relativity, clocks also slow down in stronger gravitation, so that a clock on a high-flying plane would run very slightly faster than clocks on the ground where gravity is very slightly stronger though the plane's acceleration would counteract this. The 1960 Harvard tower experiment by R.V. Pound and G. A. Rebca to test the effect of gravity on time also confirmed the theory. It is interesting to note that the forces of gravity and acceleration are equivalent in relativity theory.

We experience the feeling of weight equivalent to accelerating at g force (gravity at the surface of the earth, 9.8 m per sec per sec) upwards continually, because we are not free falling to the center of the earth. We also experience the same kind of force when we accelerate in a high-performance car or aircraft - which limits the rate of acceleration because of stress or pressure against the body of the vehicle (military aircraft, for instance).

We could not survive on the surface of a planet with a much stronger gravity than earth, and air or water vapor could not survive on a planet with much less because of the lower escape velocity.

Gravitational forces from surrounding galaxies will tend to pull a passing beam of light out of its 'straight' path thus diminishing its energy. But would increase it while the photons (light) were moving closer to the gravity source with a balanced decrease when moving away, with a net change in energy but a slight difference in direction, in principle like an asteroid looping around the sun.

Chapter 17

Black Holes Cosmic background radiation Quasars Antimatter Large hadron collider Theodor Kaluza

This phenomenon of the red shift, if it increases to infinity at a finite distance, may contribute to, or be the sole cause of the constant very low energy cosmic background radiation from all points in space detected by heat sensitive telescopes. All electromagnetic radiation from beyond the cosmic horizon must elongate exponentially (decreasing in energy) with increasing distance to infinity (or some quantum value of energy) as it reaches the exact opposite point to the point of origin.

Physicists have theorized that cosmic background radiation probably comes from the residual radiation from the original 'Big Bang' at the center of the universe. But in 4-dimensional physics there is no common center, there is no expansion and there is no need of a universal 'big bang', though there may be relatively minor explosions as astronomers frequently observe.

If the heat and light from the big bang at 7,000K (Kelvin = absolute temperature scale in which zero = -273.16C) deteriorates to near zero, why cannot light from any source in the universe do the same? There are atoms of hydrogen, plus asteroids and dust or 'dark matter' presumably, scattered through space between or in galaxies where there is very little heat from radiation. Could not these be a cause of the background radiation? Heat that bounces back and forth between low temperature and low density gas could also make a natural peak.

'Tired light' theory

There is a theory about 'tired light', which, in its very long journey, loses energy through collisions with diffused gas or dust in space. We could possibly receive light from the formation of galaxies near the cosmic horizon. Could we be receiving cosmic background microwave radiation remaining from our own region of space after it had circumnavigated the universe?

Black Holes

All visible stars are constantly converting matter to energy. In a perpetual system all radiant energy must eventually be converted back to matter to start the process again. Black holes may convert energy into new matter. Professor Feynman showed by experiment that energy has mass, so the mass of the (closed) universe is almost certainly constant. Light has mass but only because it has energy, it has no rest mass because it is never at rest. But exactly how much mass is in the form of energy, and how much in matter? And how much in dark matter or 'free' hydrogen or helium?

It seems that star systems and galactic clusters form slowly from vast clouds of dust ('dark matter') or hydrogen (the smallest atom), but how is the matter created? The biggest observed explosions occur in supernovas, but they don't have enough mass to form a galaxy. Where does this mysterious dark matter in space come from, and how is the matter formed, or reformed, from the energy constantly radiating from stars leading to their eventual 'heat death'. The dust clouds from which galaxies seem to form (they apparently are forming on a large scale) could be manufactured in mature black holes.

Yale Bulletin, 4 September 2008: Yale astronomer discovers upper mass limit for black holes. This is an observation, not a prediction. The mass limit is equal to about 10 billion stars like our sun. Hopefully, the transformation of energy back into hydrogen, or any kind of matter, which is then blasted into outer space from a black hole, keeps the system in balance.

Light must convert its energy to another form or transform to its alter ego, matter, which is known to be a highly concentrated mine of energy (as in nuclear energy). Nature abhors absolutes and infinities. This makes logical sense. Just as nature changes the rules at the quantum level, so it may at the other extreme, the black hole level. Is it possible that in this region of extreme conditions electromagnetic energy converts back to matter? What can happen to light when it strikes a black hole? Time virtually stops. Light usually converts to heat, which is kinetic (moving particles) energy normally, but a black hole isn't normal. The big question is: How does radiant energy transform back to matter? When an irresistible force meets an immovable mass what happens? The mass divides and shoots out the excess matter, apparently.

Quasars

Quasi-stellar-radio sources, or QSO: quasi-stellar object (Extracts from Wikipedia).

"A Quasar is an extremely powerful and distant 'active' galactic nucleus. They were first identified as being a high red shift source of electromagnetic energy, including radio waves and visible light that were point-like, similar to stars. There is now a consensus that a quasar is a compact region surrounding the central supermassive black hole of a galaxy."

The most luminous quasars radiate energy at a rate exceeding one trillion suns, almost equally across the spectrum, from X-rays to far infrared, with a peak in the ultra-violet optical band. By applying Hubble's law, their distances range from 780 million to 28 billion lightyears from earth. More than 10,000 quasars are known. Another source says over 200,000 are known to exist. At 28 billion lightyears this could only be explained if the redshift is caused by something other than velocity of recession, as in curved space-time (author).

At 28 billion lightyears it would take the image 28 billion years to arrive here (and more than 28 billion years for the quasar to get there), and this would only be the radius of a sphere of observable space. If the outerparts of the big bang travelled at almost light-speed the universe (if the current big-bang model is true) must be at least 56 billion years old. The lifetime of a star burning up its mass would probably be a quarter of that – another mystery of nature?

The brightest quasars devour 1000 solar masses of material every year. Quasars turn on or off depending on their surroundings. After a quasar finishes accreting gas and dust it becomes an ordinary galaxy with a black hole at its center. It should be noted that quasars that have a recessional velocity greater than the speed of light (c) are very common. Any quasar with z>1 is going away from us in excess of *c*. Early attempts to explain superlumic quasars resulted in convoluted explanations with a limit of z=2.326, or in the extreme z<2.4. z=1 means a redshift indicating travel away from us at the speed of light. The majority of quasars lie between z=2 and z=5 (Wikipedia). The mathematical sign > means 'greater than' and < means 'less than'.

It has been suggested that the excessive red shift is caused by light escaping from a strong gravitational field (a 'gravity well'), but someone calculated that the limit for this would be **a** factor of 2.4, which would not be enough for the latest figures. The earlier statement that quasars become ordinary galaxies suggests that the quasar phase takes place early in a galaxy's life.

The observation that most quasars are very distant might appear to contradict the Cosmic Principle that the distribution of matter, especially quasars, is more or less constant wherever we look, can be explained by the far greater brightness of them compared to most galaxies. The cosmic horizon is where the elongation of light waves converts them to invisible heat waves. Only the very brightest of galaxies can be seen. Therefore, close to the cosmic horizon, only quasars can be seen. Furthermore, if the elongation of light waves is exponential to infinity at a finite distance, those closest to the cosmic horizon could, with a bigger space telescope, show even greater red shifts. Quasars are relatively rare and separated by vast distances. The nearest to earth is estimated to be two million light years away.

The suggestion that quasars show that the early universe (from the big bang) was quite different from what it is now could be erroneous because, although we are looking back in time (because of the finite speed of light) they must also have spent a lot more time, at sub light speed, to get where they are now seen. If galaxies are constantly being created throughout the universe, it is likely that quasars are *young galaxies.* And if we can see more of them than ordinary galaxies it would be because they are so much brighter. "The huge luminosity of quasars results from the accretion discs of central supermassive black holes, which can convert on the order of 10% of the mass of an object into energy as compared to 0.7% for the nuclear fusion process, as in the Sun." (Wikipedia - Some technical terms deleted). There seems plenty of evidence of matter transforming to energy, but none of energy to matter, yet there seems to be no shortage of matter.

What is more to the writer's point of view is the distribution of quasars throughout the universe, bearing in mind their extreme brightness. If they were found to be distributed more or less evenly, based on probable degree of brightness at the source, not red shift, it would support the 'steady state' model of the universe.

Mysteries of the universe: Antimatter

Another mystery of the universe is the creation of matter from energy. All experiments to date have produced an equal mass, or amount, of antimatter to matter, resulting in early annihilation as the two particles touch, re-forming into energy. It is a widely held theory that for all the matter in the universe there must exist (or have existed) an equal mass of antimatter. Antimatter has opposite electrical charge to matter, otherwise it is the same (plus some other apparently unimportant (?) differences such as a slightly faster rate of radioactive decay). The theory applies to every form or particle of matter. If anti-matter decays faster would it not be a cumulative effect, eventually leaving the universe full of matter? We see no evidence of anti-matter in the universe, nor would we expect to see any. Could there be a third form of matter/energy of which we are not aware?

If the antimatter theory is correct, it is likely that it exists within the matter atom, but safely separated from it in some as yet unknown way. Where else can it be? How can we ever observe it, except for a nanosecond or two? How does natural radioactivity (the slow conversion of matter to energy) really work? It has been suggested that antimatter is involved in the tremendous output of energy in quasars. The biggest or most pressing problem in the immortal universe theory is the retransformation of energy to matter. One could assume that the energy of a nuclear bomb comes from the union of antimatter and matter. In which case antimatter exists within the atom – if the theory is correct.

The mysteries of the universe are greater than man ever thought. Scientists need to think in four (and five or more) dimensions instead of three plus a limited view of time. If we can't understand it we have to accept it in the face of hard evidence. Antimatter could exist in a higher dimension, safely separated from matter, also other force fields. One nuclear force binding particles together in an atom increases in strength like a rubber band when the particles are pulled apart!

One question about antimatter is can matter be created from nothing? If matter and antimatter are added together the result is not nothing, it is a burst of energy. The theory that for every positive thing there must be a corresponding negative thing implies that the sum of the parts is zero, but this has yet to be demonstrated. All of the foregoing indicates that matter is the universal source of energy, but none of it shows the conversion of energy back to matter – yet it must be so.

Matter or mind?

The common idea that matter is a concrete substance in the ordinary meaning of the word is being invalidated by modern discoveries. Matter turns into energy with very different properties. Space is integrated with time, forming space-time, with an amazing geometry that defies 3-dimensional geometrical logic. Every observer is always at the exact center of his observable universe, but it isn't a sphere (although it looks like one with the observer at its center), it is a continuum without beginning or end. And where does consciousness come from – not just matter and energy! At least, not without a dramatic change in properties.

We could be justified in thinking that the universe has characteristics of a construction in a great mind as well as a concrete material construction. To use an analogy, scientists argued for a long time over the nature of light: was it formed of particles or waves, because it was observed to behave like both in different experiments. The answer was both, in the sense that the waves were formed into photons, tiny quantum 'bundles or packets of energy', which behave like particles but also with the characteristics of waves. The same principle could apply to matter and mind. As the brain is regarded as the 'seat' or the physical organ of the mind, so the 'concrete' universe could be the physical organ of the mind of the Supreme Being, God.

About the Large Hadron Collider at Geneva. News item .

The Large Hadron Collider (LHC) is a gigantic scientific instrument near Geneva, where it spans the border between Switzerland and France about 100m underground. It is a particle accelerator used by physicists to study the smallest known particles – the fundamental building blocks of all things. "It will revolutionise our understanding, from the minuscule world deep within atoms to the vastness of the Universe".

"Two beams of subatomic particles called 'hadrons' – either protons or lead ions – will travel in opposite directions inside the circular accelerator, gaining energy with every lap. Physicists will use the LHC to recreate the conditions just after the Big Bang, by colliding the two beams head-on at very high energy. Teams of physicists from around the world will analyse the particles created in the collisions using special detectors in a number of experiments dedicated to the LHC.

"There are many theories as to what will result from these collisions, but what's for sure is that a brave new world of physics will emerge from the new accelerator, as knowledge in particle physics goes on to describe the workings of the Universe. For decades, the Standard Model of particle physics has served physicists well as a means of understanding the fundamental laws of Nature, but it does not tell the whole story. Only experimental data using the higher energies reached by the LHC can push knowledge forward, challenging those who seek confirmation of established knowledge, and those who dare to dream beyond the paradigm". (Internet Dec 2009)

Gravity and electromagnetism: Kaluza's theory

The mathematics of Riemann, Minkowski, Kaluza, Klein and Einstein have opened the door to multi-dimensional logic, especially Kaluza's 4th and 5th dimensional theory showing the relationship between gravity and electromagnetism, which Einstein evidently worked on for the latter part of his life without success. Although Einstein (who used many dimensions in his mathematics) favourably mentioned Kaluza's letter to him on the subject, he seems to have never mentioned it again.

There seems to be no limit to the number of extra dimensions that can be devised mathematically, but this does not mean that they can really exist. Nor does it mean that we can envisage their possibly exotic or unforeseen properties if they did. Ultimate reality might be a perfectly logical construction of many dimensions. If miracles are possible then it might well be true. Limiting the properties, or potential, of the universe, with our present knowledge, would be a sad mistake.

Black Holes

Astronomers recently announced that 26 thousand light-years from earth our Milky Way galaxy has "a massive black hole at its center called 'Sagittarius A', containing the mass equivalent to 2.6 million suns crushed into a very small volume" (New Zealand Herald Oct.15 03). It is thought that every galaxy has a similar black hole at its center. We can tell when one exists by the higher than expected speed of rotation of nearby stars. M87, a giant galaxy with an active, very bright nucleus, is thought to contain an object with a mass of about 3000 million stars.

Another N Z Herald (Dec. 9 06) report headed, 'Black hole eats star 4 billion lightyears away' confirms the possibility of regeneration. "Scientists used Nasa's Galaxy Evolution Explorer …to detect an ultraviolet flare coming from the center of a remote elliptical galaxy. This ultraviolet flare was from a star literally being ripped apart and swallowed by the black hole," said Suvi Gezari of the California Institute of Technology. "We looked at the galaxy in 2003 and there was no ultraviolet light coming from the galaxy at all, and then in 2004 we suddenly saw this very bright source." Scientists 'continue to use the telescope to observe the ultraviolet light as it fades'. When large objects approach a black hole the gravitational force is much greater on the side nearer the black hole than on the far side. Also, any centrifugal force (applied in the opposite direction) would be much greater on the opposite side than the near side.

A philosophical reflection: Rearranging molecules or electromagnetic waves into higher levels of order generally produces dramatic and useful results (as in laser beams). This should guide us into making experiments that do just this. We might discover something new and exciting. Incidentally, this also applies to thought: discipline, concentration or focus produces results.

Chapter 18

Metaphysics A general discussion Evolution Old Testament Laws

Contemplation of the existence of a Supreme Being as the all-encompassing mind or spiritual head of the universe is a natural development in the search for truth. The subject can be described as metaphysics. If there is a god ruling the universe scientists would want to know about him. Just as we have learned that there is much more to the physical universe than meets the eye, there is also plenty of evidence from the Bible, plus anecdotal evidence, that there is also spiritual mystery. The writer has to admit to having a few significant spiritual experiences that happened a long time ago, but they did not change his passion for trying to understand 'how things work'.

Such personal 'spiritual experiences' have little or no significance to anyone else, unless supported by significant events. As there is no evidence of this and the writer's history is relatively normal, he only mention them (later) in support of his beliefs, and why he feels so certain about them.

Awareness, or intelligence, could not arise from a purely physical universe of 'bricks and mortar' and electromagnetism, unless the potential for its development already existed, ready to bring forth life when conditions allowed. The question then divides into two parts: Can this happen without a God, the Supreme Being? Or is a sentient divine creator necessary?

Creation

The popular assumption is that the universe could not begin to exist without a creator. By the same logic, God could not exist without a creator, unless God were everlasting. The universe was either created by an everlasting God or there never was a beginning, and the universe is also everlasting. Anyway, if there was a time before the universe was created, what was God doing? Had he finished with a previous universe? How long did he wait between the old one and the new one?

My research shows that the universe is everlasting. God identifies himself with nature in the Bible, and nature is the perfect expression of God. A more accurate name for Nature is Nature/God, because God is the head or mind of nature, which is, or appears to be, physical with mind properties. We are products of nature and we have minds (sentience) and human feelings which are also related to the chemistry of our bodies. God and nature are 'one' as in 'mind and body'. Perhaps we shouldn't say that the universe is entirely physical because God has revealed himself to some people and left evidence of Holy Scripture. We have awareness and we are part of the universe. If we are wholly made of matter then matter is potentially conscious or sentient, or somehow linked to God.

If there is no God, the universe itself, through its grand design, produces awareness in its creations automatically, increasing the power of living things through stressful evolution until human beings (hopefully) become godlike. The alternative, that there is a God, is the one I prefer. In fact, I did say once, that if there isn't a God, there ought to be one.

I was young then and naïve. I think I must have caught God's eye because he soon set me right, and in doing so revealed that he really does exist. Why God took any notice of me at all is a mystery to me as I was (and still am) an insignificant person (see chapter 5). Man's mistake is in thinking that God's ways are like man's, reducing God to our level, but we are very different.

In the process of development nature could also, in effect, produce the likeness of an anthropomorphic God (or gods) to a large proportion of the population, but with differences of character or nature according to the times, the people, places and conditions. This would be nature acting like God. Of course, God could have created nature to be perfect like himself, so that nature would work in perfect harmony with his plan. Indications are that nature is both eternal and perfect. As God is immortal so is nature.

The existence of God requires that he is everlasting, omnipresent, omniscient and omnipotent and is perfect in his works and in his judgment. People think that nature is separate from God, because nature appears to be indifferent, without feeling. The writer believes that God and nature are one unified being. This gives us the best picture: a Supreme Being, the Head of nature, and both God and nature are perfect.

I gain much of my knowledge of God from the Bible, Old and New Testaments, but the style of my approach is that of a scientist seeking truth, which I am sure is what God would want me to do. I have long since realised that my thoughts do not originate with me but from somewhere beyond my own consciousness. I seek, and eventually I may see, but I do not know how that happens. I read the first verses in Genesis about God creating the heavens (stars in the sky) and the earth and the land and the sea, and all the creatures, and I (now) conclude that this is true in essence, because God creates and sustains all things in the universe. The older King James Version of the Bible (I prefer to quote the Revised Version because it preserves the majestic literary style of the King James) does not say that God created the 'universe', nor does the New International Version.

However, I have another intriguing view of the 'Body of God'. This is that God is the Father and that Nature is the Mother. Nature is often called 'Mother Nature'. Mother Nature is similar in some ways to a human mother, but is vastly different in others. Nature is also said to be 'red in tooth and claw' as befits savage animals; 'the female of the species is more dangerous than the male', especially in defence of their young.

Perverseness of nature

I have long noticed that there is a tendency in nature to be perverse, that nature, or fate, makes things more difficult than we normally think they need to be. For instance, it is said that 'The path of true love is never smooth'; unexpected snags appear in any undertaking. It is not scientifically provable but a female mind is often more devious, though practical and realistic, than a male mind whose idea of logic, though inventive, is often not so realistic. The female mind is frequently seen to be more intuitive and sensitive. My logic tells me that God supports 'Mother Nature' because it creates more drama, or makes life more interesting, by creating more challenges for man to overcome. Perhaps 'calamitous' might be a more accurate word than 'interesting' in acts of nature.

Also, there is a tendency for the poorest countries (and parents) to have the most children (because only two or three will probably survive to become parents themselves), and for the richest to have the least (because of a high level of survival). Italy, not long ago had a high birthrate, but now has a very low birthrate. Male logic might expect well-to-do people to have the most children and the less successful (financially) to have fewer, but the opposite appears to be true.

The rules of logic that govern the universe might explain some things. Just when we have to do something personally very important, some of us are inclined to feel very nervous (fear of failure?) and this can spoil our effort. Wanting something very much can build up nervous tension that defeats our object. Trying to hide something from observers can draw unwanted attention and so on. Mother Nature can be very trying to many people but overcoming obstacles probably has its advantages. Fate, or the future, is unpredictable in the sense that often the 'best laid plans go awry'.

Man's emergence

The writer believes that man emerged from the animal level with a singularly clear conscience and closeness to God.

Once they realized they had the objectivity and intellect to follow several paths of behavior apart from instinct they fell into temptation becoming degenerate and hopelessly benighted. It was all foreseeable. This explains the existence of Adam and Eve in Paradise representing the dawn of the age of mankind and man's subsequent fall. The New Testament shows man the way to regain Paradise lost. There is no division in heaven.

"The sons of God saw the daughters of men that they were fair: and they took them wives of all they chose. When they bore children to them, the same became mighty men which were of old, men of renown (excerpts from Genesis 6. 2&4)." I believe it is possible that the 'sons of God' were the earliest uncorrupted generation of men, and the 'wives' were from the later generations.

Man is adaptive

Scientists say that man is not descended from apes, but that we are both descended from an unknown or unspecified common ancestor. Man developed intelligence and speech relatively quickly because he firstly developed tool-making hands. Tree-dwellers developed hands for grasping branches. Man developed an opposing thumb making holding things easier. Weapons would soon have followed. The amazing thing is that man is very adaptive. We must adapt to whatever is necessary or perish. In other words, learn. Man's superiority to animals comes from learning from others, and from previous generations from sophisticated speech and writing. Also from the acceptance of civilized and cooperative behavior.

If God created man

I suggest that, if God created Adam and Eve as literally described in Genesis, he would have created them to be perfect, and they would not have fallen into temptation. There would have been no need for a savior, no struggle for survival and enlightenment, and no need for faith. In other words, there would have been no drama. It might have been basically unjust in principle, apparently because there should be no reward without effort. Reality, compared to a dream, has its price. We all need drama. "Man shall not live by bread alone, but by every word of God" (Luke 4.4).

God adapts his style of revelation to the people and the times. God presents himself to those people as being in the likeness of a man: an anthropomorphic God to whom they can relate and in the kind of language they can understand. God is also anthropomorphic to me but with infinitely greater attributes: literally omniscience, omnipotence, and an ethical standard (based on logic) that would be superhuman, hence our difficulty in interpreting the 'ethics of nature' or apparent lack of them.

No difference between nature and God

Without going into detail, the literal interpretation of creation in six days about six thousand years ago, as described in the Old Testament, is not the way of the 'universal' God as I have observed all my life. I find no difference between nature and God. I do not find nature doing ungodly things, nor God doing unnatural things. He has no need except for special purposes. Nature is far more complex than we realize and is perfect – like

God. If we look at the broad sweep of nature rather than details it is easier to see.

If the reader disagrees with this widesweeping statement we must consider that God is just and cannot err. Everything is logical, even morality or ethics. The logic of man is of a far lower order than God's perfect logic. God is not just intelligent, he is omniscient. He is also omnipotent, which means that what might appear unjust to us can become just within God's plan. Untimely death is as we see it. If it is God's will, it could be the entry into heaven. Suffering must have its purpose, or fulfil several purposes, one of which is the motivation to avoid it and another might be to 'purify' or make us more humble (make us more savable).

'Intelligent design'

This 'observation of nature' is a highly subjective observation and cannot be proved, but it was enough to convince me that there is a universal God. He also overrides local Gods, his local revelations as God to specific peoples. There is a mind behind nature. I suppose the latest idea of 'Intelligent Design' in nature, even though it cannot be proved to the satisfaction of some people, is correct. I strongly doubt that anyone will believe this, or see as I do, but I am certain that it is true.

Will of God

I respect the Old Testament and the revelation of Jehovah to Israel, but I have some reservations about the authenticity of some (or some parts of) other revelations which seem to follow the human cultural style, at least in parts, rather than the divine style. I believe that heaven cannot be divided against itself and that God, who cannot sin or err, is justified in all things. Where there is a difference between revelations (as in obvious contradictions) I am sure there is a reason, such as error or corruption. Corruption is likely where scriptures are in the hands of corrupt rulers. The end result will be the will of God.

We have to remember that God is responsible for the whole universe, whereas man is only responsible for his personal 'space' and responsibilities. Furthermore God is responsible for both the present and eternity, and for the efficient development and replenishment of the universe. A universe that is not logical would become chaotic. The universe has to provide challenges and the opportunity for fulfilment of all living things *forever.* Nobody wants to be bored. Even animals in zoos can die of boredom.

The universe is designed to provide a constantly changing environment. Our tiny minds cannot compare to the infinite intelligence that guides the universe. Only divine inspiration can give us the knowledge of God, and then only that which we require in God's judgment to fulfil his will. It is said that the only 'constant' in the universe is change. 'God moves in mysterious ways his wonders to perform.'

Age of the eartholy Spirit of God.

Some of the available evidence of the great age of the earth exists in the presence of coal and oil deep in the earth formed from trees etc., which were once on the surface. The layers of thousands of fossils in rocks at different levels formed over millions of years show the past presence of life. The beds of rivers such as the Grand Canyon gorge, 2,000 to 6,000 feet deep (about 3 and 1/3 feet equal a meter), which must have taken millions of years (17 million is the most recent estimate) to create, all invalidate the idea that earth was formed in six of our days. The age of the earth is the same as the sun and is estimated to be 4.5 billion years old, and didn't have any oxygen in its atmosphere until life forms extracted it, some microbial forms of which still exist today.

As we drill deeper into the earth the temperature rises showing that the earth is slowly cooling from the center outwards, unless natural radioactivity is renewing heat. The interior is molten rock, and nearer the center molten iron (creating the earth's magnetic field) and nearer still to the center (the writer suggests) very heavy metals such as gold. Note that things get lighter in weight as we approach the center because of the mass above. Planets, and moons commonly surrounding the bigger planets, indicate the probability that all suns (or stars) probably have planets like the sun.

God's days

The Old Testament is valuable, but it must be understood in context to the times and conditions of its revelation. To take it as the literal Word of God in modern times at the same level and significance as the New Testament of Jesus is counterproductive and confusing. The Bible does not save; Jesus does, and Jesus did not write the Old Testament, Jehovah, the Jewish (revelation of) God did. There is only one God but many revelations that differ relatively to the then current circumstances. God does not necessarily act in the way that human beings think he should.

The story of creation in Genesis is a poetical account of an underlying truth, that God creates all things and that nature is the expression of God. Also, God's use of the word 'days' is for man in a term that primitive man could understand, not the four billion years or so that the earth has been in existence.

Insisting that Genesis is a literal account of creation is trying to deny a self-evident truth, that the world is more like four billion years old. Trying to subvert education in science to teach that the world was created about six thousand years ago is illogical. Why do they want to do this when God creates worlds like ours continuously? Cannot they imagine that God's domain is the whole universe? Too much knowledge of science apparently transfers the wonder of creation from God to nature. Is this a test of our faith? If God exists he controls nature throughout the universe. We just have to readjust our idea of God from an earth God to a universal God. Jesus didn't talk about science, he taught people how to behave and think, and how to become closer to God. That is more important to people than part-knowledge about science.

Perhaps they are thinking of their followers, that they will not understand and fall away, throwing out the baby with the bath water. People have knowledge today of science that would have been impossible to make sense of to people even two thousand years ago. The writer seems to have the same problem with people today, trying to convince them of something unfamiliar and mystifying to them that will probably become common knowledge in the future.

Of course, this is asking us to believe that man is the product of evolution: merely beasts with bigger brains. Well, there is plenty of evidence that we are, both genetically and in our 'average' (with wide variations) lifestyles. One has only to listen to the world news to hear of widespread brutality. Our bodily design shows remnants of animal design that we have not yet lost. It is supposed to be the business of religion to raise our vision and aspirations above the level of animals. God, through nature, created life on earth that evolves into many forms but with one physical principle: survival through reproduction and continual adaptation. What is the point of arguing about *how* God works? Let us give thanks that there is a God.

Man the scientist is also the expression of God whose infinite knowledge includes the future, so that his purpose unfolds as he wills. False theories of science and false prophets are all instrumental for God's ultimate purpose relatively to us, which is the raising of the best human beings to the heavenly estate, and the destruction of the worst. Only God knows the way that God chooses to do this. The writer believes that God cannot sin, which means that God is justified in all that he does. If we had enough knowledge we could understand this in detail, but with present knowledge we can only understand it in principle.

Subversion of education

Insisting that the Genesis account is literally correct, instead of correct in principle, is not helping the image of the church. It is not unlikely that they uphold the literal truth of the Bible for their followers' sake, or they don't think it is an important issue, and it may do more harm than good to discuss it. Some people like things to be black or white and are intolerant of shades of difference. But life is not black and white; there are shades of difference. The importance of the issue has been magnified by trying to subvert education in colleges, no doubt with good intentions. But good intentions often have severe consequences because of underlying faults, and have been the cause of great destruction. There is a saying that the path to hell is paved with good intentions. If these people will persist that the Old Testament is infallible or immutable why don't they still stone people to death for infractions of 'the law'?

Logic of salvation

Elsewhere I have described the logic justifying salvation. By forgiving others God is justified in forgiving us, and so on. We do not kill people for being unbelievers. In older, more harsh times, the revelation of the 'universal' God in the name of Jehovah gave a set of laws in response to that society's spirit or culture. The Old Testament puts the fear of God into us. We need it, even if we are Christian in belief. One attempt to remove the Old Testament in a Christian church resulted in a revolt by the congregation.

'Immutable word of God'

It irritates me when religious people deny the self-evident truths of scientific discovery. It destroys their credibility. Their excuse for doing this, upholding the whole Bible's literal credibility - the immutable word of God - does not make sense. Jesus changed the teaching in the Old Testament in several ways: diet, law (eg. death for adultery), keeping the Sabbath (which Sabbath?), divorce (Mt 19. 3-9), religious practices such as the ritual washing of hands (Mt 15 1-20) and, most notably, sentiment.

Jesus continued in Mt. 15. 17-19: "Do you not see that whatever goes into the mouth passes into the stomach, and so passes on? But what comes out of the mouth proceeds from the heart, and this defiles a man. For out of the heart come evil thoughts, murder, adultery, fornication, theft, false witness, slander. These are what defile a man; but to eat with unwashed hands does not defile a man." Ritual hand washing was practised by clerical Pharisees who did not perform physical labor.

With regard to hand washing, in a specific test, more germs were passed onto equipment in surgery after the hands were washed than before. This was because the hands were not properly dried. We only have to touch the handle of a door to be recontaminated. Jesus objected to compulsory ritual hand washing. Some germs are not necessarily bad in a world full of seriously bad germs as they can stimulate the body to keep up its resistance to disease.

From John Stott's book, 'The Contemporary Christian' (previously quoted) "We Christians cannot surrender either the finality or the uniqueness of Jesus Christ. There is simply no one else like him; his incarnation, atonement and resurrection have no parallels. In consequence, he is the one and only mediator between God and the human race. This exclusive affirmation is strongly, even bitterly, resented. It is regarded by many as intolerably intolerant. Yet the claims of truth compel us to maintain it, however much offence it may cause (p.64).

Jesus appears to supplant God, but only because he was sent by God for God's own purpose. He was only obeying God so the argument goes to God. Why did God have to upset the previous and obvious (to man) course which was directly between God and man? Jesus answers the question best himself. Briefly, Jesus is God's justification to save. Belief in Jesus is the crucial test. Many believe in God, but few believe in Jesus because they see him only as a man, yet God has given him dominion over the earth with power to give life.

The time and setting were not ready for the revelation of God through Jesus Christ. Even now most of the descendants of 'Father Abraham' (Jews and Arabs) reject Jesus. Islam in particular is very defensive regarding Christianity. Jesus was a descendant of Abraham through his mother so he was half-brother to his mother's other children (Mt.12 46). There might be a traditional enmity between Jews and Muslims because Sarah was the mother of Jews and Hagar was the mother of Arabs who eventually became Muslims.

Chapter 19

Can different religions live in peace? The growing power of Islam Art and religion Kosovo Other beliefs Atheism, a belief that God doesn't exist.

The modern view among leading Christians and people who don't have strong religious views is that all religions can be tolerant of each other and live in peace. So they can if their culture and Holy Scriptures allow. Some parts of the Koran sound very peaceful but other parts are not. Regardless of whose fault it is that conflicts exist, the fact is that they do, they have done and it virtually certain that, given time, they will do. We can only judge by history, ancient and modern. Christians have recently been cruelly persecuted in parts of Indonesia, which is set to become a wholly Muslim country.

The most important difference between Islam and other religions is that Islam is also a political system, a theocracy with Sharia law with harsh, rigidly enforced rules that reduce women to second class status compared to men. This has not emerged in European countries because it would be ineffective as a minority. But even as a minority Muslims tend to be prominent because they think, dress and behave differently. The adults probably can't change having been indoctrinated with culture and religion since birth, but the children, once they have grown up, might leave home and choose their own life (if the females are allowed without a great fight with the male members of their family). The power of Islam comes from sternly enforced traditional culture, which favors male dominance over females within the family, supported by the Islamic religion.

The growing importance of Islam in the West

Almost a quarter of the world's 6.8 billion population is Muslim, according to a new global survey. Two thirds of all Muslims live in ten countries: Indonesia, Pakistan, India, Bangladesh, Iran, Turkey, Egypt, Algeria, Morocco and Nigeria. Europe is home to about 38 million Muslims, or about 5% of its population. Germany appears to have more than 4 million Muslims. In France the overall numbers were lower but the percentage of population was higher. Of 4.6 million Muslims in the Americas more than half live in the United States. In the Western provinces of China serious rioting broke out between minority Muslim Uighurs and majority Han Chinese. (From AP article printed in NZ Herald Oct 10 09.) Muslim families traditionally have 6.5 children.

The practical differences between Christians and Muslims are mainly in their traditions or culture. At present there are some differences that are incompatible, particularly regarding the treatment, status and freedom of women. The so-called circumcision, but often mutilation, of female children is regarded by the west as particularly obnoxious, is not mentioned in the Koran, is condemned by the World Health Organization and is now unlawful in some countries including Egypt. The best we can do is 'agree to disagree', but agree to keep the laws of the land in which we live. In the meantime keep abreast of present day happenings in world news between Muslims and Christians to learn facts rather than pious opinions. Mohammed (570-632), founder of Islam, although he had several wives after his first wife died, had no surviving children except a daughter. Biographies are available on the Internet.

The Kosovo war allegedly happened because Christians couldn't live with Muslims and felt forced to leave over a long period. The Kosovo war was particularly vicious. European governments took the side of the Muslims, probably for pragmatic reasons. In the Lebanon war in which the European governments became involved the Israelis handed over their defensive positions to the Lebanese Christians. Although most of Macedonia's ethnic Albanian minority are Muslim, they have generally been secular, but radical Muslims have recently rebelled against the government over this issue.

Islam is fast becoming the most powerful religious and political force in Asia through to Africa and showing growth throughout the western world including America and Europe. Islamic religion reinforces the Arabian culture, which hasn't changed for two thousand years, except that they worship one God instead of many. It puts the teachers of Islam in a strong position in regions with a similar culture. In other words, it puts the stamp of Koranic approval on a culture that is already in existence. Is this argument substantially correct? Islamic culture is man-centered with man next to God. Heaven is described as a beautiful garden (Paradise) with 72 beautiful virgins waiting for the young men who die giving their lives fighting for Allah, but some interpreters of ancient Arabic (which is difficult to read because of the absence of vowels or marks and difficulty with reading some consonantal lines) say that the word for 'virgins' (hur) really means a particular delicious variety of grape. Christianity teaches humility and service to humanity. We do not know what heaven is like because we do not know the capacity or talents of the 'elect', except that 'the least in the kingdom of God is greater than the greatest among men born of women' and there will be no marriage.

The writer would not have mentioned Islam were it not an important religion affecting Christians, if not now then at some time in the future. Muslims are noted to respect power, because they believe that power comes from God. Hence, the use of power is not a sin in their eyes, especially for religious enforcement for which they expect God's reward. They seem to assume that God approves of Arab culture and that it should apply everywhere. We never hear of imams preaching peace within Arab nations. Indeed, they probably can't for fear of consequences. Muslims believe Allah is God. The 'Universal God' of the writer's description has no respect for any religions; he simply judges on the facts without favors to anyone. As we behave towards others so will he treat us, as Jesus taught. "Vengeance is mine, saith the Lord."

The Universal God is primarily a God of justice. He becomes a God of Mercy only if we deserve it by being merciful to others. He is compassionate if we are compassionate. He reveals the truth if we seek it with a humble heart, not with biased preconceptions.

Justification of the Cross

To repeat the essential message about Jesus: Jesus is the Light of the world. He suffered and died on the Cross to justify the salvation of sinners. The prophet Mohammed clearly did not think that God needs justification. But how can a just God save sinners? Sinners cannot be saved without justification. Uninformed people would not know this; they think God can act on a whim (like man) without justification. But that is not the nature of God.

Where does the will of man fit in the Christian faith? Will is in second place to judgment. A strong will is worse than useless if it is misdirected. Strength of will is important in Christianity as in any religion or lifestyle. Determination and perseverance, in other words 'high quality concentrated effort' is necessary to achieve anything worthwhile. The 'Will of God' is the ultimate expression. Jesus must have had the will of God to achieve divinity, or it was predestined.

The Old Testament was written before Jesus appeared so Jesus was not mentioned except in prophecy as the Messiah. But God's promises of salvation, the same in principle as Jesus' promises, will be fulfilled according to the same rules: Love God and your neighbor.

We know that everybody will not become a Christian and, in the Christian tradition, we do not judge the person for not believing, only God can do that. God has all knowledge but we have little knowledge. Perhaps a few remarks on other religions besides Islam might keep things in perspective.

Other beliefs

Buddhism is more of a philosophy than a religion. Buddha was asked, "Is there a god?" to which he (allegedly) never replied. I understand they have a very peaceful code of behavior, but the essence of their beliefs is that man should strive for a state of desirelessness, for desire leads to suffering. The peace of desirelessness is also Nirvana, a 'blissful state' (dict). I perceive it as a merging of self with 'mystical nature', beyond desire. Also, that reincarnation allows man to do better in subsequent lives until he achieves Nirvana. But what if he doesn't improve, or gets worse? Presumably he will suffer hell on earth in this life and in subsequent reincarnations. The writer can't understand the purpose of everlasting suffering. but that is an area out of his reach.

Resisting temptation, which is sinful or ungodly desire, that is selecting good and rejecting bad desires is within the Christian religion. Rejecting all desire, even the scent of flowers as in pure Buddhism is probably the best philosophy that man could produce omitting God. The 'Great Buddha', a famous statue of Buddha, was destroyed by the Taleban recently because it was considered an idol, and therefore an offence to God.

"Contrary to popular belief, there is only one god in Hinduism. However, this one god (referred to as Brahman) is expressed in multiple ways. Nirguna Brahman is God without form, or God without personal attributes. All personal forms of God in Hinduism, such as Vishnu or Shiva, are different parts of

God in personal form or God with attributes" from Wikipedia, Internet. The religious part of Hinduism seems concerned with raising the level of consciousness through yoga until divinity or Nirvana is attained.

I suggest that the reader consults Wikipedia or experts on the Internet concerning other religions from both opposers and defenders to get a balanced view. A reliable opinion cannot be obtained without reliable facts. It sounds reasonable that we can all live in harmony but modern history shows that some religions and cultures need a change of philosophy to achieve this.

The Cross and the Sword do not sit well together and medieval culture does not mix well with modern culture. I suggest that culture changes with level of affluence and education. The EEC governments seems to think that immigrants will eventually adopt prevailing customs but they are proving very resistant and provoking a ground swell of discontent, especially in Switzerland. It seems that affluence is not the only factor governing culture. Islam is so encompassing that it must be very difficult for Muslims to change any aspect of its culture without a feeing of guilt. Apostasy or conversion to another religion is widely regarded as treason and can be punishable by death. It is important to the establishment (the ruling class) that every person is fully committed to Islam less division weakens authority and the entire nation. Islam is not without its critics within its borders but they don't usually survive for long.

Art and religion

Middle Eastern Islamic art is restricted to geometrical patterns so they do not give offence to God; refer to the Old Testament of the Bible, Exodus 20. 4-6: "You shall not make for yourself a graven image, or any likeness of anything that is in heaven above, or that is in the earth beneath, or that is in the water under the earth; you shall not bow down to them or worship them or serve them, for I the Lord your God am a jealous God, visiting the iniquity of the fathers upon the children to the third and fourth generation of those who hate me, but showing steadfast love to thousands of those who love me and keep my commandments." Muslims interpret this literally, which is the reason for the predominance of geometrical patterns in the Middle East.

Does this mean we can't paint portraits, beautiful or dramatic scenes or erect statues to famous people? Or does it mean just that we shouldn't worship these examples of human endeavour? Commonsense would suggest the latter. A lot of religious paintings glorify God. When religion makes people behave badly there is something wrong with religion or its interpretation. Anyway, people do not worship idols or statues these days. They would think it was ridiculous. Jesus taught us to pray to our heavenly Father, and we usually add 'through Jesus Christ'. Some people pray to and through their favourite saints but this is a personal preference in my opinion. The important part of prayer is the sentiment or depth of feeling associated with it. Prayer itself is an act of worship. To pray for something unacceptable to God is a sign of ignorance or wrong teaching as Jesus warned. If we pray for something for which we can do something we should do it, then God might reward us for our effort.

Atheism

At the present time atheism is being vigorously promoted and receiving some support. The contention is that atheists are as good (and sometimes better) than many Christians. The simple answer to this proposition is: why not become a Christian and receive the gift of salvation (we all need it) and eternity in heaven. It is not logical to assert that God does not exist unless we have facts to support it, only insufficient acceptable evidence for committed atheists to believe what they want. There is a lot more evidence to believe in God than not believe, and a very big difference in consequences. Until a person has experienced closeness to death's door I doubt if anyone can be sure he doesn't believe in God or hell. 'A little learning is dangerous', get more knowledge to be sure. As a Christian a man has much to gain in this life and the life to come, and nothing to lose except his guilt. Enlightened atheists rely on intuition (as most of us do) and Western culture, which is largely founded on Christian morality.

Atheists blame religion for much of the trouble in the world, but the real cause is man's ignorance and sinfulness. Condemning all religions is just as extremist and aggressive as the worst religions.

Chapter 20

Communion with Christ His ascension and divinity Sentiment St Paul The hard commandment Christian teaching literally universal The Archangel Gabriel

There is no better thing that man can do than aspire to divinity, communion with God, which we can gain by communion with Christ. As God is the head (in complete control) of nature (as the body, or expression, of God) so Jesus gained complete control of his body and mind, in communion with God, so that he could save mankind. Jesus has perfect communion with God, and mankind is offered communion with Jesus, through faith and the holy sacrament of bread and wine. The Christian Church is defined as the 'body of Christ'.

The power of Jesus to perform miracles comes from his power over his 'spiritually extended' body, nature, which is also God's 'body'. God the Father and God the Son are 'One', as Jesus said. Divinity, in any form, has communion (one body) through the Father who has power over all. Jesus, who claimed pre-existence as the Son (second person in the Holy Trinity) of God, referred to his human self as the 'Son of Man', born of woman and the Holy Spirit. Thus, Jesus would know his power over nature and, to our surprise, power to raise his own body from death. The Holy Spirit of Jesus is the same Holy Spirit of God. The title, 'Son of man' would mean 'son of mankind' and 'Son of God', which is how Jesus often referred to himself. Also he referred to 'Son of man' in the future.

The words 'Son' and 'Father' are human words used to convey human meanings that may not adequately describe these relationships in heaven. Jesus said, "The third day I shall be perfected" (Luke 13. 32); "Touch me not, for I am not yet ascended to my Father." (Jn 20. 17). I do not think we can describe or even imagine the exact relationship between heaven, the abode of spiritual beings who can appear in our world at will, and earth, nor how the resurrection will be accomplished. It isn't essential knowledge, but the future heaven does seem to involve a merging of the two.

Any divine incarnation would have similar inspiration and power to perform miracles that Jesus had. Saints, mortal human beings with an exceptional degree of holiness, would have some degree of spiritual power to heal and perform miracles at appropriate times, and inspire others to believe in Jesus. The physical world and the spiritual world are designed to work in perfect harmony. We can never understand everything in the physical world, but we can understand the essentials of the spiritual world through which we can control the physical world. Jesus gave us clear instructions, not to gain power but to be made divine.

When Jesus spoke to a sick man saying, "Be healed" or "rise and walk", he would not know in medical detail how this would be done, or how it could be accomplished instantly. He would know he had the power, but someone had to understand the meaning intended and have the knowledge and power to do it. That could only be God. It would only be God that put the opportunity, the thought and the action into effect.

He spoke of evil spirits possessing people and cast them out. It was effective and made sense in the context of the time. What was the reality, or the real explanation? No one knows. Perhaps the spirits were real, or they represented a method of understanding (code or language) something that could not otherwise be understood. Miracles would not be miracles if they could be understood. If Jesus were here today he would probably do the same things in the same way.

Divinity of Jesus

He redefined the image of God, which scandalised the teachers of religious law of those days. That, his claim to divinity as the Son of God (which they did not understand), his criticism of their unbelief, his alarming power of healing and growing popularity made them call for his death at the hands of the Roman governor. A god cannot be killed. The resurrection of Jesus was proof of his divinity. Jesus lives in heaven and will come back to earth in the last days. "I will not drink of the fruit of the vine, until the kingdom of God shall come (Like 22. 18)." This last statement implies an earthly Paradise.

Compare the Old Testament, Numbers 15. 32-36 where a man was stoned to death for gathering sticks on the Sabbath, with Mt 12 9-14 where Jesus, in defiance of the Pharisees in the synagogue, healed a man's withered hand, and said, "So it is lawful to do good on the Sabbath." Also, from Luke 14. 3-6: "And Jesus answering the lawyers and Pharisees, saying, 'Is it lawful to heal on the Sabbath day?' And they held their peace. And Jesus took the man with the dropsy and healed him. And answered them, saying, 'Which of you shall have an ass or an ox fallen into a pit, and will not straightaway pull him out on the Sabbath day?' And they could not answer him again on these things." The Old Testament appears to be inconsistent in the matter of punishment. All sinners are not treated with the same severity. Had they been, all would have perished.

Sentiment

The sentiment in the Old Testament emphasised the severity and fear of God with harsh punishments for breaking divine laws. In the New Testament the emphasis is on forgiveness, the love of God and salvation through Jesus. St Paul (formerly a strict Jew) made his judgment very clear on Mosaic Law: there is no salvation in the law, only through faith in Jesus.

He expressed his opinion of Jewish law, or rather its lack of effectiveness regarding salvation as opposed to faith: "I through the law am dead to the law, that I might live unto God...I do not frustrate the grace of God; if righteousness comes by the law, then Christ is dead in vain." (Gal 2. 19-21). In the Old Testament God gave the Jews his laws; in the New he gave the world Jesus Christ for our enlightenment and salvation.

Is Jehovah a God of justice or a loving God?

The way the world is created he appears to the writer to be both. He is punitive towards the unjust and unrighteous, but is loving towards those who are loving towards people and God as Jesus taught. If the wicked prosper, it is worldly prosperity and it is only in this life. God is omniscient and he must have his reasons. The reward of the faithful followers of Christ is life everlasting in the kingdom of God. Jesus teaches about rewards in the first few verses in Mt. 6.

God is patient in his judgments. If he wasn't patient everybody would be afraid to sin because they would be fully aware of the consequences (as in Islam, the Old Testament and harsh civil law and order). Life would not be a test of man's true feelings, his faith and perception. Instead, God lets nature (apparently) deal with the problem of correction. Although the wicked person thinks he sees clearly, or what is obvious, he does not think of the future, of heaven or hell or God, because he has no faith. He doesn't believe that nature is perfectly logical and that consequences eventually match causes in the highest sense. Or if he thinks of God at all he rationalizes (finds excuses) or falsely justifies himself in his judgments, perhaps thinking he is doing good. Societies where much of the populace is virtually lawless usually fall into one of the two extremes: severe control of wrong doers, according to the standards or beliefs of the rulers, or virtually no control at all as long as the main rulers are not affected. Commonsense and common humanity would provide correction of a sensible order to minor offenders increasing for second offenders until it is clear that only long sentences (or separation) will provide safety to the public. There is no reason why such people should not work to help pay for the expense they cause. There is always much work to be done. There should not be any unemployment problem in a well-organized country.

The hard commandment

Jesus repeated the commandments to the young rich man who wanted to know what he needed to do to be saved: "You know the commandments: Do not commit adultery; Do not kill; Do not steal; Do not bear false witness; Defraud not; Honor your father and mother" (Mark 10. 19). "All these have I observed from my youth" he replied. Then Jesus beholding him loved him, and said: "One thing you lack, go and sell whatever you have and give to the poor; and you shall have treasure in heaven; and come, take up the Cross, and follow me." *Jesus was pressed* to give the perfect answer in the circumstances, but it would not have been the perfect answer except for a person seeking perfect faith or a person who already has great faith who would know or understand what he was about to do. In any case Jesus was there, with his disciples, for him to follow.

A person with perfect faith would have to be a perfect person, an incarnation in human form with divine knowledge and power such as in the kingdom of heaven in the likeness of Jesus. A divine person is not an independent person, standing alone in an indifferent world; he is a finite incarnation of an infinite God who also controls every aspect of the environment, from the smallest space to the whole universe. He is inspired, guided and protected by an omniscient, loving and omnipotent God through the Holy Spirit. But what mortal man can elevate himself to that divine estate? That is why we need a Savior.

When the rich man turned away he commented to his disciples about riches, then rephrased it with, "It is hard for them that put their *trust* in riches to enter the kingdom of God." Money is not evil, it is the love of money that is "the root of all evil". In *coveting* money "they have erred from the faith, and pierced themselves through with many sorrows" (St Paul, 1 Tim 6. 10). A perfect person would have no need of money. He would be divinity incarnate, the expression ('Word') of God. A perfect person would not try to 'tempt' God because his motivation would be wrong and therefore a sin. Inspiration comes to us as we need it or when we are ready for it. A perfect person would be a god. Ancient Greeks wrote about gods, but they were not perfect.

Making a god of Mammon (money) is a cause of temptation. Our first duty is to God, who then rewards us. To expect a reward robs a virtuous act of its merit. It is what we do with money that is the critical question. Philanthropists try to do the most good with it according to their opinion or their advisers.

Jesus warned us not to tempt God, such as throwing yourself off a cliff to prove that God will save you (ref Mt. 4. 6). Also, it presumably takes time to grow into such a holy estate that you know that what you are about to do is both right, and within your capability. I doubt that Jesus had to practice to perform miracles; he would have known by divine intuition or knowledge that he had the power.

Strict God

The writer emphasizes that a strict God is essential for certain people as the power behind Jesus to enforce what Jesus taught. The laws of the universal God are strict and cannot be broken without penalty, but the penalty is applied by God himself at the time, place and in a manner according to his judgment. Likewise, God rewards our virtues. If the writer's attitude to the Old Testament seems cavalier it is because, in the writer's opinion, it needs to be read with understanding of the times and conditions. Besides, parts of the Old Testament if they are interpreted literally, are frequently criticised by unbelievers to the detriment of the New Testament by association.

Genesis is notable for its picturesque descriptions of Paradise: the creation of man and woman, their temptation by a serpent, the tree of knowledge of good and evil, the tree of life by which man can live for ever, and the creation of earth and sky in seven days; all of which can be explained as a poetical description of an underlying truth. There is much wisdom in the Old Testament. The fear of God is the beginning of wisdom.

Jesus questioned

Again, on the question of divorce, the Pharisees, testing Jesus, said to him, "Why then did Moses command one to give a certificate of divorce, and to put her away?" He said to them, "For your hardness of heart Moses allowed you to divorce your wives, but from the beginning it was not so. And I say to you: whoever divorces his wife except for unchastity, and marries another, commits adultery." (Mt 19 7-9). Verses 3-6 give a fuller explanation, in which husband and wife become one when joined by God. Even some of the disciples were not happy to hear this (about divorce). Jesus added another somewhat mystical explanation concerning celibacy (v 12), "for the sake of the kingdom of heaven. He who is able to receive this, let him receive it". Saints, who are also mystics, intuitively would understand this. Self-control is part of divine discipline, raising us to a higher level of awareness.

Heavy emphasis on keeping laws without understanding when they become inappropriate to the real circumstances makes them an injustice, and the opposite of the intention of the law, which should be to serve man, not man to serve the law. Cultures where the people are rebellious and ungodly by nature receive both government and divine revelation that are harsh, frightening and uncompromising to set an example. Such revelations are 'just' when viewed in the context of the times and the people, but they are not designed or meant to directly lead to salvation as we understand it, although they may have other merits such as developing strengths that are stepping stones to a higher revelation.

It is very clear in the Old Testament that Jehovah made himself the personal God of Israel, and I can understand the reluctance of Jews to look elsewhere. But they cannot have it both ways. Jesus came later with a new revelation that, in some ways, altered the old one. Local religions reflect local cultures. The 'Word of God cannot ever be wrong' assertion has to be understood to be moderated by the language and circumstances, which can vary greatly. The simple thoughts of man cannot describe the knowledge of God.

The words of God depend on the person or people he is talking to. Abraham was praised and rewarded for his faith and certain virtues, not for his Christian virtues. By the standards of his time he was a great man; by modern civilized standards he would have been severely criticized, as would King David who won praise for his power and success rather than his morals. The writer's 'Universal God' is, by definition, the only true God, and all other Gods are representations of him to suit the environment of the time. There is only one God and his rule is universal regardless of local manifestations. One has only to interpret history and observe nature and people to know this. I should add that I believe that Jesus is the true prophet and incarnation as the Son (appropriate finite presence) of the 'universal' God.

There is a message for Jews in the New Testament (Mt.15.22-28) concerning the Canaanite woman whose daughter was severely possessed by a demon in which Jesus says, "I was sent only to the lost sheep of the house of Israel." Why did he say 'lost sheep'? But when he saw her humility and great faith he healed her daughter instantly. There is another account in Jn 10 16: "And I have other sheep, that are not of this fold; I must bring them also, and they will heed my voice. So there shall be one flock, one shepherd. For this reason I lay down my life, that I may take it again. No one takes it from me, but I lay it down of my own accord. I have power to lay it down, and power to take it again; this charge I have received of my Father." Who are these 'other sheep'? Are they potential Christians who have never heard of Jesus?

Christian teaching universal

By accepting that God takes into account the different levels of development and kinds of personalities of the recipients in his revelations, we preserve the concept of a just and impartial God, and the genuineness of the recipients. There is no doubt that the teaching of Jesus is the way to salvation and God, because it is rational and supported by the mass of evidence in the New Testament. It is also literally universal in that it would be true for any advanced society anywhere.

The Christian doctrine depends on the immanence of a loving and just God who controls everything, keeping order in the universe, dispensing justice with a far-sighted strategy that accomplishes his divine will.

We cannot 'judge' people because we are not omniscient. Only God can properly do that. We may not judge what they truly 'are', not knowing their past experiences and upbringing, but we can certainly be offended at what they do. I feel duty bound to enlighten anyone who will listen, and all the enlightened should do likewise. "Do not hide your light in a cellar, that those who enter may see the light. Your eye [view] is the lamp of your body; when your eye is sound your whole body is full of light; but when it is not sound your whole body is full of darkness. Therefore be careful lest the light in you be darkness." (Luke 11. 33-35).

Jesus gave his disciples a commandment, "Preach throughout the world." Missionaries going to foreign lands where they are not welcome and badly treated should note that Jesus also said, "If anyone will not receive you or listen to your words, shake off the dust from your feet as you leave that house or town. Verily I say unto you, it shall be more tolerable for the land of Sodom and Gomorrah in the Day of Judgment than for that city" (Mt 10. 14, 15).

The writer quotes the words of a refugee from Somalia: "We nearly all come from a collectivist culture. It is like we are one body, and can only function effectively when all our parts and all our processes work properly. In individualist cultures, the emphasis is on each person being independent – functioning fully on their own." In a free society we must respect the right to disagree, but it would be a crime to resort to violence over differences of religion or culture. What does a collectivist culture imply? Does it mean that all should follow the community views? I think it should mean to be as agreeable as possible without offending your conscience, otherwise it means that cultures cannot mix without trouble. Either we agree to live in peace or keep separate. It depends on how incompatible the cultures are.

The angel Gabriel

The Archangel Gabriel appeared to Mary the mother of Jesus, telling her that the child she would bear would be holy and "shall be called the Son of God" (Luke 1. 35). Gabriel also told her about the time of birth of her cousin's child who would become John the Baptist. Is this the same Archangel Gabriel that appears in the Koran that contradicts what he said to Mary the mother of Jesus the Son of God and miracle worker? From the Internet:

Qur'anic statements which portray Christians and Jews in a negative image (9.30, 5.72, 3.85, 4.150, 58.22) include verse 30 of Al-Tawba which states:

"And the Jews say: Uzair is the Son of God; and the Christians say: The Messiah is the Son of God; these are the words of their mouths; they imitate the saying of those who disbelieved before; may God destroy them; how they are turned away!"

"And verily We gave unto Moses the Scripture and We caused a train of messengers to follow after him, and We gave unto Jesus, son of Mary, clear proofs (of Allah's sovereignty), and We supported him with the Holy spirit. Is it ever so, that, when there cometh unto you a messenger (from Allah) with that which ye yourselves desire not, ye grow arrogant, and some ye disbelieve and some ye slay?"(Koran, verse 002 087). Note: Muslims do not understand the Trinitarian Doctrine of God. They place human limits on God who is omnipotent. They (or some) apparently regard Mary as the Holy Spirit. They speak of Jesus speaking comforting words to his mother the moment he was born. They do not believe in his crucifixion, asserting that the authorities were tricked and that Jesus went to another country – all this to explain why he could not be the Savior and Son of God that he claimed to be (author). On the other hand Jesus is sometimes honored and say his teaching should be believed.

To illustrate the 'Fatherhood' or Superiority of God, regarding the time of Judgment Day, Jesus said that no one knows the time except the Father. Jesus is not competing with God. Imagine if one 'normal' person had all knowledge of past, present and future, how could he 'live' or interact with mankind? How could he remain interested in what he already knew? Only by identifying or empathizing with living beings could he live vicariously in the sense that we live. Reversing our position we have faith in God's care, that as long as we fulfill righteousness we shall receive God's blessing and eventually become divine with appropriate powers and knowledge to maximize our potential to enjoy life fully in company with others in the Kingdom of God. But how do we fulfill righteousness? How do we know what to do and what not to do and be sure that we follow the right teacher or prophet? We should all do some thinking and research for ourselves and weigh the evidence and the kind of people who claim to be preaching the word of God, or interpreting the teaching. If you are good you will be attracted to what is good and you will become better. If you are full of fault you will be attracted to evil, unless you are humble enough to admit that you are sinful. Then God (by any name) will help you to become more pure (free from the bondage of sin) until you are raised and fit to enter the Kingdom of God. Study the New Testament and trust Jesus, and do what is within your capability. Do not trust an unchristian culture if you can avoid it. God wants us to feel responsible for ourselves and for others. Personally, the writer distrusts cultist supposedly Christian religions in which women do not wear modern clothes and do not mix with ordinary decent folk. Normal good behavior separates people naturally. It doesn't require us to be outwardly or ostentatiously different. As the saying goes, "Birds of feather flock together".

Responses

The main Quranic statement that Muslim scholars portray shows Christians and Jews in a positive image in Surah Baqarah, Chapter No. 2, Verse No. 62:

"Those who believe, the Jews, Christians and Sabians - any who believe in God and the Last Day, and work righteousness, shall have their reward with their Lord. They need not fear, nor shall they grieve." Not only is this paragraph an enlightened statement, it is clear and poetically written – a rarity in all respects in the older versions of the Koran in the writer's opinion.

Because the Koran leaves itself open to different interpretations I borrowed an old copy from the local library and finally found the passage: "Believers, Jews, Christians, and Sabaeans – whoever believes in Allah and the Last Day and does what is right – shall be rewarded by their Lord; they have nothing to fear or to regret." (Translation by N. J. Dawood. First published by Penguin Books 1956, latest (revised) edition 1978 (quoted).

By changing one word 'Allah' to 'God' (Referring to 'God' in the Old Testament) as in the Bible and writing in a superior style the Koran has been made very much more acceptable, even praiseworthy. Unfortunately Allah's 'We' who do the speaking to the prophet Mohammed in the name of Allah give a completely different account of the birth of Jesus. In his Introduction Mr Dawood writes, "God (for Allah speaking to Mohammed in his visitations through the angel Gabriel as 'We') speaks in the first person plural, which often changes to the first person singular or the third person singular in the course of the same sentence." Islam accuses Jews of corrupting the scriptures, and

Christians of worshipping Jesus as the Son of God instead of God, which Allah (if he were God) would surely know was not true, but Mohammed would not. Why is God called 'Jehovah', or 'Yahweh' in the Old Testament and not 'Allah'?

The writer's impression is that Mohammed, who received these visitations sometimes in 'trances', or at least one in his dreams, which he remembered vividly, influenced or imposed his own level of understanding on the messages possibly subconsciously rejecting what he disbelieved in relation to the lifestyle or culture around him, but that is only an opinion. There is no proof or confirmation of the source of his message except that the literary style is claimed to be poetic in the original Kufic script that has no vowels or diacritical points. The writer's experience of reading the Koran, after a little while, with the frequent repetition of the theme "In the Name of Allah, the Compassionate, the Merciful" was that it had a hypnotic effect, which he didn't like, and found contrary to the sentiment of the main message.

His message passed through many hands of possibly talented scribes who might have rewritten the thoughts expressed in a more presentable way than the original. Parts of the Koran use long sentences and other parts short ones. There was no apparent chronological order. Mohammed could see the angel Gabriel, but his daughter, who was with him sometimes (or at least once) said she could not see him, saying, "Who are you talking to Papa?"

Mohammed could neither read not write, and his background was commercial as a caravan trader working for a wealthy widow 15 years his senior, who became his first wife. When he was asked for guidance on punishment for an adulterer he called for the 'Book' and based his answer on Mosaic Law. Consciously or subconsciously he wanted a religion that was special to Arabs. He made drinking alcohol a sin, prayer three or five times a day, fasting during the month of Ramadan, a pilgrimage to Mecca, halal killing of meat by cutting the throat and draining the blood with the animal facing east or Mecca, pig meat forbidden as unclean, women to be ruled by a man (father, husband or brothers) and so on. As a caravan trader he was ruthless on thieves (cutting off hands). He rejected the Christian message probably because be barely understood it as is the case with most Muslims today.

Some of these rules have serious inconveniences in a modern industrial mixed society, especially frequent prayers. Whatever he thought he was successful in his cause, but that does not mean that it was right in the abstract sense. It has been successful mainly by force and the sword and expansion of population, reinforced by heavy indoctrination and prohibiting women a proper education. Now it is being seen as a serious threat by its opposers. Alternative beliefs are regarded as hostile and dangerous to Islam and Allah. Their very existence is regarded as an affront to Allah. Christians and Jews, reasonably, have as much right to be offended by Islam.

Allah, through Mohammed, has usurped or supplanted God as presented in the Old Testament and rewritten it to favor Islamic sentiment with his own version of events so that Abraham, Moses and other prophets are recorded as worshippers of Allah a thousand or more years after the original version. Where the Bible says Jehovah or God, the Koran says Allah (or God, now for the benefit of

Westerners.) We should call Islam's Holy Book the Revised Version of the Bible by Allah through his Archangel Gabriel (expressing himself as 'we') to the prophet Mohammed. If it was Allah that spoke to Moses, why did not Moses record that fact?

In the Koran every chapter is preceded by "In the Name of Allah, the Compassionate, the Merciful", and then condemns all unbelievers to the horrors of everlasting hell. For instance: "The Day of Judgment is the appointed time for all. On that day no man shall help his friend; none shall be helped save those on whom Allah will have mercy. He is the Mighty One, the Merciful. The fruit of the Zaqqum-tree shall be the sinner's food. Like dregs of oil, like scalding water, it shall simmer in his belly. A voice will cry: 'Seize him and drag him into the depth of Hell. Then pour out boiling water over his head, saying: "Taste this, illustrious and honourable man! This is the punishment which you doubted." "As for the righteous, they shall dwell in peace together amidst gardens and fountains, arrayed in rich silks and fine brocade. Yes, and We shall wed them to dark-eyed houris. Secure against all ills, they shall call for every kind of fruit; and, having died once, they shall dic no morc. Your Lord will through His mercy shield them from the scourge of Hell. That will be the supreme triumph". (Pages 148,149; v. 44-34, 36, 47).

All readers should note that the Devil is traditionally a tempter and a deceiver. He tempts people with the things they desire with promises that will certainly not be fulfilled. To be fair to Mohammed his 'religious experiences' seem similar in style to Moses. God speaks to Moses as God; Allah speaks to Mohammed through two angels as 'We' confirming much that Moses reported to his people. The name 'Allah' predates

Mohammed as one of the many gods worshipped in Arabia up to Mohammed's time. The revolutionary idea of one God was introduced by Mohammed to his people. It gave an increased sense of unity that increases strength. He also gave them a religious code of behavior that was harsh on women, adulterers and thieves, and demanding in religious duties. He could see drunkenness had bad effects on people so he banned the consumption of alcohol. He unified Arabia and strengthened religious conviction, making them a serious military force with a hatred for anyone not like themselves.

Where are the critics in Islamic countries? Who can question the authenticity or veracity of the Koran, or parts of it? Who can analyze it and discuss it except the clergy who are already committed to it and dependent on it for their livelihood and, perhaps, their lives? The clergy say that the only way to understand the Koran properly is in the original Arabic. The true inspirer who changes the Bible appears to the writer to be the enemy of the God of the Bible. It couldn't be God because he would be contradicting himself, or he might have allowed it for reasons known only to himself. Islam claims that Jews and Christians corrupted Holy Scripture, but why should they? What profit is there in corruption? Why not follow the truth? For that matter why didn't Allah follow extant Holy Scripture instead of changing it, unless he had other plans, like the salvation of all Muslims and, by implication, condemnation of all Christians, Jews and unbelievers in Allah and his prophet Mohammed?

It all comes down to belief as well as reality as in good deeds and bad deeds (as judged by Allah). Non-Mohammedans now have two spiritual enemies if Allah is correct: The Devil and

Allah. If the foregoing is confusing think of the poor believers faced with a diabolical choice, everlasting hell if they make the wrong choice, or Allah's heaven for men with billions of virgins (because there must be billions of male Muslims). The writer has no idea what awaits Muslim women, because it's unlikely that children will be conceived in heaven, unless there is no sexual difference and all will be 'as angels' as Jesus said.

In the Christian heaven 'there will be no marriage, nor giving in marriage' but 'the least in the Kingdom of God will be greater than John the Baptist'. It is likely that there will be no sexual differences in heaven or the Kingdom of God. What these 'beings' will do will depend on their powers or capacity, but they will be immortal and raised to divinity. The greater the capacity, physical or mental, the greater the sphere of interest.

The important question is whether *all* Muslims will follow the peaceful injunctions or some the aggressive ones. We can only judge how dangerous they are, not by their words but by their deeds. The world is terrified of Moslems who seem bent on persecuting or murdering critics, both within and without Islam, and nobody can be sure of where the next terrorist attack will come from. Universities who take pride in fearless research in particular are restricted. One professor in an Islamic country was thrown off a balcony by his students. What makes the situation worse is that the background of Islam is relatively shaky and its prophet's biography is worldly, not as other prophets. Search the Internet: 'Koran criticism'. Just for interest, Allah says that thieves should have their hands cut off, but Muslims say that they don't do it these days. Apparently the clergy can change Allah's rules if they want' What are Allah's rules about that?

If the writer offends people by using the word 'shaky' it was meant to remind followers that Islam is founded on the words of Mohammed from a spiritual being that no one could see, a message that has been passed through many hands of scribes and political authorities, and that apparently contains contradictions according to different interpretations that are too serious to overlook. If people choose Mohammed in preference to Jesus, it can only be because the message of Mohammed is preferred to the teaching of Jesus. They are free to chose as we are.

All belief that is based on faith alone should be understood to be unjustifiable in civil law, unenforceable and private, particularly when some laws are self-evidently unjust. Civil manners to everybody are required in a civilised society otherwise, in these days, we cannot live in peace. People are moving in their thousands from poor Islamic countries to wealthy (or wealthier) western countries with a mainly Christian culture. Without change, tolerance or compromise there will be increasing friction especially with extremist Muslims.

Extremists who preach and practice extreme indiscriminate violence are like a cancerous growth in the body of Islam, the likely consequence of which is death of the whole body. Is this the image that Allah would wish for his worshippers? The Koran follows the sentiment in the Old Testament and some parts of the New Testament, but not much, especially not the divinity of Jesus. Violence by Muslims against Christians in some Muslim countries is almost a daily occurrence as reported in Church publications and world news.

Nearly all of the Koran is based on stories already extant in the Middle East at that time according to scholars and seem related to the likely opinions of the prophet Mohammed and his later followers, considering his lifestyle, as he later became, a very successful and feared warlord. His thoughts, or memories of the revelations, were passed on orally. They were later written down fragmentally and later still collected and put into order. They resulted in different versions of which one was selected and the rest destroyed. In other words the authorities decided, or were tempted to decide what people should hear.

The reason for their extraordinary success is their appeal to men in that they support their culture of male superiority or lordship next to Allah, and conquest. To repeat the writer's description of the universal God, he gives to people what they justly deserve. If God were not justified he would not do it. The fact that he sent a Savior, Jesus, and gave him power to save and judgment over all people on earth, means that God sees merit in some of earth's people. If the writer is correct, Jesus will judge according to the reality, furthermore, his judgment might reflect the person's judgment of Jesus, for that must have a great bearing on that person's behavior, which is the reality.

If Jesus is what he claimed, and all the evidence confirms what he said, what does that imply for his unbelievers? Who is traditionally the greatest enemy of God? The Devil is supposed to be the Archenemy, the 'Antichrist', who deceives mankind if he can, but there is no need for a spiritual tempter as there is plenty of natural temptation. So it does not matter whether we believe in the Devil or not. But it is a good idea to assume that he could exist because it consolidates the whole idea of temptation in a malevolent spiritual being, which is truly how temptation seems to work. Unfortunately, the writer cannot embrace the idea himself because he cannot imagine anyone, with the Devil's alleged cleverness, stupid enough to fight God, except men, and then only in ignorance – "forgive them for they know not what they do". To justify the Devil's existence he would have to be an angel serving the Universal God acting as the Devil to test mankind. Sin corrupts mind and body, so it is important for prophets to be pure, or as pure as they can be.

In the writer's considered opinion the prophet Mohammed was a deep thinker and could see the need for a suitable religion in his region, which at that time was composed of many gods (of which the most, or one of the most, important was called 'Allah). He could see the superiority of a one-god religion, especially its unification of the followers into one powerful body. According to his biographers he was subject to 'seizures' in which he saw the archangel Gabriel who spoke to him as 'We', but no one else could see him.

Followers of Jesus drink the sacrament of Holy Communion in which a small quantity of red wine represents the blood (life) of Jesus, and eat a small cube or wafer of bread, his body. This is an outward and visible public acknowledgment or witness of an inward and spiritual grace, that we are (or assent to be) one (mystical) body in Jesus as he commanded (Mt 26. 26-29). Jesus has knowledge of God, and he and God are one. Jesus proved it by his miracles, divine teaching and his resurrection.

Chapter 21

Three Old Testament prophesies confirming Jesus as the Messiah St Paul The archangel Gabriel Baptizm Sabbath God is responsible for all of creation Virgin births

Isaiah 9. 6-7: For unto us a child is born; and the government shall be upon his shoulder; and his name shall be called Wonderful, Counselor, The mighty God, The everlasting Father, The Prince of Peace. Of the increase of his government and peace there shall be no end, upon the throne of David, and upon his kingdom, to order it, and to establish it with judgment and with justice from henceforth even forever. The zeal of the Lord of hosts will perform this.

Micah 5. 2: But you, Bethlehem Ephratah, though you be little among the thousands of Judah, yet out of you shall come forth unto me he that is to be ruler in Israel; whose goings forth have been from of old, from everlasting.

Daniel 7. 13-14: I saw in the night visions, and behold, one like the Son of man came with the clouds of heaven, and came to the Ancient of days, and they brought him near before him. And there was given him dominion, and glory, and a kingdom, that all people, nations and languages should serve him; his dominion is an everlasting dominion, which shall not pass away and his kingdom shall not be destroyed.

To confirm in the New Testament the statements written in the Old Testament: Rev. 1. 17-18: When I saw him, I fell at his feet as though dead. And he laid his right hand on me saying, "Fear not; I am the first and the last; I am he that lives and was dead; and behold, I am alive for evermore, Amen; and have the keys of hell and of death."

"The Old Testament is supremely the book of expectation. From the start, people are shown as living under hope of a promised deliverer. This promise narrowed and clarified in the course of history. The promise was first concentrated in the seed of Abraham. Then, later, the promise was associated with the royal house of David. Meantime, Moses himself linked the promise with the notion of a coming Prophet. Thus, from many main angles of Israelite life, a 'Messianic' expectation came into focus (P32, New Bible Commentary, Third Edition, Inter-Varsity Press 1970)."

The titles applied to Christ are only a few of the many ascribed to Him in Scripture. Among the others are (Isa 40. 9; Jn 20. 28), the Almighty (Rev 1. 8), the bread of life (Jn 6. 35), good shepherd (Jn 10. 14), Lord of glory (1 Cor 2. 8.), King of kings (Rev 17. 14), and Lamb (Rev 13. 8).

JOHN 8. 54-59: Jesus replied, "If I glorify myself, my glory means nothing. My Father, whom you claim as your God, is the one who glorifies me. Though you do not know him, I know him. If I said I did not, I would be a liar like you, but I do know him and keep his word. Your father Abraham rejoiced at the thought of seeing my day; he saw it and was glad." "You are not yet fifty years old," the Jews said to him, "and you have seen Abraham!" "I tell you the truth," Jesus answered, "before Abraham was born, I am!" At this, they picked up stones to stone him, but Jesus hid himself, slipping away from the temple grounds.

The Old Testament does not mention the Holy Trinity, but it does mention God's Spirit several times. Nor does it mention God's Son, but it does refer to 'Son of man' who has the status and power of the Son of God. Jesus spoke of himself as the Son of man. He also spoke of God as 'my Father', and told us to pray to God as 'Our Father in heaven'. He never said, "Pray to me". Christians, at the end of a prayer, usually say, "In Jesus' name, Amen".

God the Father, Son and Holy Spirit

The Holy Trinity concept of God is too sophisticated for most people (perhaps everybody) to fully understand even in these days, so it was probably not appropriate at all to use in the Old Testament. In the New Testament the Holy Spirit gives power and knowledge (communicates, gives gifts and executes God's will). Jesus said, "The Counselor, the Holy Spirit, whom the Father will send in my name, he will teach you all things, and bring to your remembrance all that I have said to you (Jn 14. 26). We have to remember the limitation of words in any language to convey the intended meaning accurately, also that words can change their meaning over time with common use. The Church teaches that Jesus was conceived by the Holy Spirit, and born of the Virgin Mary, as described in the New Testament by the Archangel Gabriel. On earth he was justified in saying that God was his Father. Speaking of his risen state he said, "Before Abraham was, I am". Thus 'Son' would not be the literally correct title in the universal view. The writer describes Jesus as the finite Presence of the infinite Father.

If God is everlasting so is his 'Son', his offspring and likeness in a localized form for the benefit of man, full of wisdom and the knowledge of God. People judge Jesus. Those who deny his divinity, and say that he was just a good man but was not whom he claimed to be, may be judged themselves with the same judgment. Those who believe may be saved by the same logic. I think we underestimate the power and importance of faith. God is just. An omnipresent Supreme Being would not present himself as a finite being in one place at one time (such as us) without changing himself and assuming an appropriate form. The finite divine being (presence) has divine powers and knowledge appropriate to his purpose. His purpose would always be appropriate to the circumstances because all would be the will of God. Remember that God controls the listener as well as the speaker, and that Jesus is the highest prophet of God. Jesus came to earth in his human form both to teach and to test us.

St Paul's religious experience

One of the most important incidents that show the divinity of Jesus is the miraculous transformation of Saul, the ardent Jew, to the most ardent and gifted disciple of Christ: "I am verily a

Jew, born in Tarsus, and brought up at the feet of Gamaliel, and taught according to the perfect manner of the law of the fathers, and was zealous toward God, as you all are this day.

"And I persecuted this way unto the death, binding and delivering into prisons both men and women. As also the high priest doth bear me witness, and all of the establishment of the elders from whom I also received letters to go to Damascus and bring them that were bound to Jerusalem to be punished.

"As I made my journey and came near to Damascus about noon, suddenly there shone from heaven a great light round about me. I fell to the ground, and heard a voice saying to me, 'Saul, Saul, why do you persecute me?' And I answered, 'Who are you Lord?' And he said to me, 'I am Jesus of Nazareth, whom you persecute.' And they that were with me indeed saw the light and were afraid, but they heard not the voice of he that spoke to me.

"And I said, 'What shall I do, Lord? And the Lord said to me, 'Arise, and go into Damascus and you will be told of all things which are appointed for you to do.' And when I could not see for the glory of that light, being led by the hand of them that were with me, I came into Damascus. A devout man called Ananias said to Saul, 'receive your sight' and Saul was able to see again. Saul was also called Paul and used the name Paul from about that time. He had a turbulent history as he preached without fear in a dangerous period. His writings form about half of the New Testament after the four Gospels.

We don't know what they think, but we know what they do.

I have no 'social' problem with people's beliefs as long as they act in a civilized way – which I define as the Christian way, with kindness, consideration and understanding. We cannot see inside people's minds, we can only see what they do. People who are by nature kind, compassionate, loving, honest, cooperative, helpful, humble, not self-righteous or self-opinionated or judgmental etc are both civilized and Christian in deed if not in faith. If they do not believe, it could only be that they do not understand; they have not been enlightened. If they are truth-seekers rather than being self-assertive, ie 'knocking on the door' it should be opened to them.

Why a Savior is necessary is beyond our comprehension in detail, but the simple answer is that it justifies God, whose righteousness (as in logic in scientific thinking) is perfect. How, otherwise, can a just God justify saving sinners, or in saving some sinners and not others? We are all sinners, both in what we do and what we fail to do. Human rationality is usually pathetically inadequate, and worse when they are obviously opinionated without good reason.

One of the things I am trying to do is break the barrier to belief that has arisen from science, which deals in physical things and, by implication, ignores spiritual things, especially belief in God, and the need of a divine Savior to justify salvation. A person without faith is not likely to rise above sin. We cannot please God without faith (Heb. 11 6).

If a person is seeking faith "but cannot find it, ask the question: what is wrong with me that God has not granted me faith`? Love God; God *is* Love, and love your neighbor as yourself. It seems that God is in us when we do godly things. We should do godly things all the time, and never ungodly things, then God will dwell within us.

My contention is that a personal God (a revelation of the universal God) reflecting the character of the revelation to specific persons, regions or races, must change the revelation to suit the new times and conditions; especially when there has been much intermarriage with other races as is evident with the Jewish nation. The character of the Jewish nation in the new Israel is different from the old Israel. It is certainly less bloodthirsty, and they have been less aggressive than they could have been. Also, they enjoy more freedom and a higher standard of living and education than two thousand years ago.

There is only one God, but many divine revelations, ranging from the lowest to the highest. Ancient religions can be regarded as appropriate revelations. They may have serious differences but they were, and could still be, appropriate to certain people, even if it means that, in the opinion of Christians, they are misled. God gives intuitive knowledge, conscience and judgment, as well as laws, which if people disregard makes them sinners.

To do evil in the name of the law is a sin. It is in the nature of written laws to fall short of righteousness, or always be appropriate in a multitude of different situations. Likewise, laws may be absent when they should be present, as many people have found when they sought redress for harm done to them. Personal judgment is sometimes required to interpret and possibly override inadequate or inappropriate written law.

Jesus spoke of God as his Father, and to pray to God as 'Our Father in heaven'. He implied, by repeating some of the basic Ten Commandments (without the penalties), that Jehovah was the same God (plus other references). This is not in disagreement with what I am saying. There is only one God everywhere, but the messages or revelations he sends to different peoples, on the basis of divine justice and reason, may be very different.

The fact that this may lead to wholesale condemnation - which happens - is all part of God's judgment. God is judging all the time. Satan (or evil people) may do the actual work (in Christian theology) but God, being omnipotent, takes personal responsibility. We also feel responsible for our own actions, which must be part of God's will. God's conversation with Satan concerning Job makes Satan appear a servant rather than a dangerous enemy. At least he is, perhaps, an essential part of the divine scheme. Temptation is certainly real, so is the hidden deceit associated with it.

These statements might offend people, but God is not a human being with human duties and obligations. He cannot sin nor do anything unrighteous. He is justified in everything that he does or causes to be done. It is not for us to do the things that God does. God is omnipotent and omniscient. If he wanted everyone to be saved in the Christian sense he would do it, and he could do it now. What is right to us as Christians and aspiring children of God is only a small part of God's will. God is active everywhere, doing those things that only God can do.

God takes responsibility for all of creation. He is omniscient and judges correctly, which we cannot do. God is continually fulfilling his plan. He requires us to have faith, not require proof of his existence. This is God's way, and there must be very good reason for the way he acts. Suffering and death is necessary in this world. It is not for us to inflict it. It is God's prerogative, which he does through his chosen means.

In the writer's philosophy, death is the judgment of God through nature. It is the natural consequence of inherent sinfulness and ignorance. By ignorance I mean the lack of enlightenment which comes only from God. By enlightenment I mean knowing how to think, speak and act (and how not to think, speak and act) that is both virtuous and righteous, and pleasing to God. To possess perfect enlightenment is to be divinity incarnate, a representation on earth of God the Father in heaven.

At all times the universal God rules, which we witness in nature. Man is prophesied to almost be destroyed. The power of Jesus is the power to save, so that sinful man can survive and be raised to divinity. The life of Jesus is an act of God, and Jesus is the 'local expression' of God, because God and Jesus are also 'one'. A better understanding of this can be gained by reading Jesus' own words in the New Testament.

The Virgin Birth of Jesus

The word 'Son' has a human meaning that the child will grow up to become the likeness of his father, with inherited characteristics of both parents. It does not follow that the life of Jesus, begotten of God and the Virgin Mary is identical to the human experience. It does not mean that there are two Almighty Fathers. Jesus made it very clear that he obeyed God, and that his teaching was from God. He said, "Do not call me good; only God is good." (Mt 19 17). He ascended into heaven and has the power of God over earth. He has all the divine attributes necessary to do the Father's will, which means he has perfect communion with the Father. He has fulfilled all that the Father sent him to do. Whoever does not believe in the Virgin Birth of Jesus, and that he is the Son of God, does him an injustice.

The angel Gabriel was sent by God to a city of Galilee named Nazareth to the Virgin Mary, saying, "Fear not, Mary, for you have found favor with God, and you shall conceive in your womb, and bring forth a son, and shall call his name Jesus." Then said Mary, "how shall this be, seeing I know not a man?" Gabriel said, "The Holy Ghost shall come upon you, and the power of the Highest shall overshadow you; therefore that holy child that shall be born of you shall be called the Son of God. He shall reign over the house of Jacob forever, and of his kingdom there shall be no end".

As this last prophecy has not yet been fulfilled, it does not necessarily mean that Jesus is not the Messiah of the Jews. Jesus, as the Messiah, took everybody by surprise by appearing to be humble and (apparently) quite human, going to his death on the Cross. His resurrection, as he predicted, proved he was all he said he was, and more. At his Second Coming who knows how, or when, he will arrive? God seems to like surprising us. The Bible gives several descriptions. One of the great lessons of the Bible is that we are required to have faith unto death. By faith we are saved.

It might be of interest that there have been several reports of virgin births but they were all female babies. This, apparently, is in accord with scientific theory. I believe the report and can see the logic of the theory that virgin births must be female. However, I argue that this supports the belief that Jesus was unique, and exactly what he claimed, because he was born male. Not only that but he knew who he was from an early age. His whole life was planned by God, and Jesus knew it.

Baptizm

After his resurrection Jesus told his disciples to meet him at Galilee. "And Jesus came and said to them, 'All authority in heaven and on earth has been given to me. Go therefore and make disciples of all nations baptizing them in the name of the Father and of the Son and of the Holy Spirit, teaching them all that I have commanded you; and lo, I am with you always to the close of the age' (Mt 28. 18-20)." From Jn 20. 21, 22: "Peace be with you. As the Father has sent me, even so I send you. And when he had said this, he breathed on them, and said, "Receive the Holy Spirit."

There are many disputes about baptism, about what it means and how to apply it. Is infant baptism (followed by confirmation) correct or must it be when the person is mature and understands the sacrament. Is full immersion necessary? The writer endorses infant baptism because it confirms the parents' wish that it follows Christ for the rest of its life whether it is confirmed later or not (confirmation apparently has been discontinued). It is also a beautiful and comforting ceremony for the parents. A child of more mature years may still not have the maturity to make a decision or commitment for himself.

Claims are made that baptism is being 'born again' with a Christlike character and life ever afterwards. Unfortunately, this is self-evidently not true. A leopard does not change its spots overnight though, in some cases, where there has been a sudden change in beliefs or philosophy there can be a distinct improvement. Many people are told to expect a miraculous change, but are disappointed. Many strong believers display unchristian attitudes in daily relationships showing their ignorance. Only in the kingdom of heaven will people become divine.

What did Jesus really mean? In simple terms: become believers and followers. Sacraments, such as baptism and Holy Communion, have their place and purpose; but they are public outward and visible signs of what is hopefully an inward and spiritual grace. How, in detail, they are performed seems relatively unimportant as long as they are fitting to the circumstances. The spectacle of seeing someone climbing into an ugly zinc bath to be fully immersed seems in bad taste. Jesus did not baptize anyone with water, but he did by the Holy Spirit. He instructed his disciples to baptize but he didn't say how. John the Baptist introduced baptism, the idea of which he probably got from a similar purification ceremony prevalent at the time. But he did introduce or spread from the Essenes the idea of repentance for sins, which was new and became very popular. Baptizm is the act of receiving the Holy Spirit by any acceptable means according to God's will. True baptizm means a public commitment to Jesus Christ. This is the reality.

Baptizm was a symbolic cleansing ceremony but repentance was the essential and practical requirement. Repentance is, or should be, a progressive state. Overcoming one worrying sin is likely to reveal a previously hidden one and so on until we are perfect, divine and immortal, but no one has ever succeeded except Jesus who overcame 'the world'.

There might be good reasons why a person is not publicly baptized, in which case commitment in the heart and mind would surely be acceptable to God who is omniscient. There would be many that have been baptized who are not committed or who have lost their faith. Unless there are good reasons to do otherwise, the writer supports the commandment of Jesus to be baptized, and eat the bread and drink the wine of Holy Communion at reasonable intervals (say once a month) in honor of Jesus who said, "Take, eat; this is my body." And he took a cup, and when he had given thanks he gave it to them, saying, "Drink of it, all of you; for this is my blood of the covenant, which is poured out for many for the forgiveness of sins. I tell you I shall not drink again of this fruit of the vine until that day when I drink it new with you in my Father's kingdom." It is the correct thing to do.

The writer's personal style is to make his religion invisible, but behaving as well as possible. True religion should be natural and constant, in harmony with God and nature. However, for good reasons, the writer has never preached or been involved in church activities. Instead, he practised Christian living within his ability as any average person would. In this way he was able to test and research Christianity in a practical way in his own life without any bias. He felt that the knowledge gained would be more valuable than becoming openly religious. What was Jesus doing in the first thirty years of his life? With his power and knowledge as the apparent son of a carpenter with very normal half brothers he kept quiet until he was ready (or God was ready) to begin his difficult three year ministry. The writer only mentions this to show he is not the first person to adopt this unusual policy and that many may do it to some extent. Probably every individual has his own style of worship, if that is the correct word.

Sabbath

In the Christian world the Sabbath was changed from the Old Testament Saturday to Sunday to honour Jesus. Where he alters the laws in the Old Testament, the New Testament takes precedence. Jesus is not a man, full of fault. He was, is, and always will be, the divine Word (finite presence and expression) of the infinite God. God is infinitely greater than the personal tribal God that the Jews thought he was. He is the God of the universe, who has given power and judgment over the earth to Jesus. Had Jesus not come into the world, the Old Testament would probably be virtually unknown, except to Jews.

The Sabbath, or seventh day of rest, would have been arbitrarily chosen by existing custom, since days existed long before man. The essence of the purpose for the commandment would be to ensure that people wisely kept one day out of seven to rest and worship God. I do not think that God ever rests or sleeps. It is good that we should worship God. It is in our human nature to require frequent inspiration, and it pleases God that we should put temptation behind us and seek godliness through Jesus.

The Sabbath was moved to Sunday from Saturday to commemorate the resurrection of Jesus. "For as the Father raises the dead and gives them life, so also the Son gives life to whom he will. The Father judges no one but has given all judgment to the Son, that all may honor the Son, even as they honor the Father. He who does not honor the Son does not honor the Father who sent him. Truly, truly. I say to you, he who hears my word and believes him who sent me, has eternal life; he does not come into judgment, but has passed from death to life" (Jn 5. 21-24).

Jesus performed miracles and healed the sick as an example to his followers, to glorify God, to draw congregations and to spread the word of salvation. Christianity is an educational process. We have to do our best to purify ourselves, raise our standards and expand our minds, and improve our behavior, thoughtfulness and manners, especially in day to day things, which is within our capacity.

Just as a mental exercise, suppose Jesus cured everyone's sickness in one moment, nobody would die and the population of the world would increase alarmingly with all kinds of trouble. The time and place for that is in the resurrection. The healing Jesus performed was mainly for proof of his divinity and for our example and inspiration.

It is the small things that reveal a Christian's true nature and level of understanding. Our pre-existing ideas, convictions and habits do not usually change overnight. Enlightenment does not come easily or quickly. A perfect Christian would be Christ incarnate, immortal and divine. We should experience, over time, an increase in our empathy (notably lacking in serious sinners), an elevation in perceptiveness and judgment, and an increase in faith and spiritual insight; in other words, an elevation of our entire personality.

Chapter 22

The Lord's Prayer Theological leadership The 'dead' Infinite incarnations Simplified Unitarian view of God Fatalism

There is a section in the 'Lord's Prayer' (Mt 6 9-15) which does not seem right to say: 'And lead us not into temptation'. It has been changed in recent times to: 'Save us from the time of trial'. I only mention this as the original form confirms the previous statement to Moses that God takes responsibility for everything, in particular, things that we would assign to nature. It might seem insignificant, but coming from such an important utterance, and from Jesus, I regard it as significant. The Devil (ever the Deceiver) tempts, but that is only part of the divine scheme. Temptation is usually natural to human nature but all temptation is ungodly desire regardless of its cause. In overcoming temptation we grow stronger.

By overcoming all temptation (which can be difficult to recognize) we become perfect. Think of temptation as a test of faith and willpower, and to make us stronger in character and mind. Without an evil force to resist how can we become strong? Life is full of challenges, and temptation in its many forms, for instance, to surrender to the temptation to 'give up', is part of all of them.

The universe would be like a body without a head if God did not exist. Therefore, if God is everlasting, so is the universe, and God has power to change it, if he wants. He is the Lord God Almighty in the literal sense of the word. His creation of living beings must give him satisfaction. We might never know how close he is to us unless, or until, we become divine.

Lord Jesus

A god-man (such as Jesus) who has perfect communion with God speaks as God with divine authority. Although the Father is always greater than his revealed local presence, he dwells in the divine being and shares in his life. Jesus suffered on the Cross but the Father did not. Using the analogy of the human being the brain 'knows' pain but apparently cannot experience pain itself. Physical injury to the brain itself does not cause pain. Headaches occur near the surface of the head and transmitted by nerves.

God is spirit and has no form. The principle of relativity (the point of view of the observer) applies to our relationship to God. We can all communicate, or commune, with God at the same time. Our life, or consciousness, is part of God, because nothing can exist without God. Therefore, a divine being incarnate is justified in experiencing the knowledge of his divinity, which is supported by his power over nature. It is therefore correct to call Jesus 'Lord', in the sense that he is the physical and spiritual presence of the Almighty Father whose domain (or body) is the whole universe. There is perfect harmony in heaven.

Fatalism

It would not be wise to be fatalistic (because God controls all things) to the point of not trying to do what we should be doing, because this would diminish our character, and be a direct cause of the very things that we would want not to happen. We feel, and are held responsible for things that are in our power. Rationalization is not reliable because we are not omniscient. We should not judge ourselves, nor make limiting assumptions about God.

Infinite incarnations

The nature of God, a Trinity of states with three separate functions and persons, yet all forming one Almighty God allows for infinite incarnations. In the kingdom of God there will be many such incarnations, all divine and immortal. The Unitarian, single entity, idea of God places a human limit on God's attributes or Being. It is a great folly to place any limit to God's power or form. The Trinitarian concept of God is a divine revelation.

Simplified view of God

Some churches teach a simplified view of God (and the Bible) because they feel that they would lose their flock if theology became too hard to believe or understand. Theology is important to thinkers who like to understand, while more practical people want practical advice. Jesus gave practical advice, but it is surprising to the writer how many Christians prefer their own (unchristian) opinion when it suits them. We should remember Jesus' warnings on this subject. It is intellectually very simple to be a good Christian: just believe in the Lord Jesus.

Theology

But belief implies a lot more because we have to learn from the New Testament what Jesus taught. Clever people often do their own thinking about the theology of Christianity and do themselves (and others) a lot of harm. Much of the intellectual leadership of the established churches has little or no faith. It is the leaders of society that set the trend, so it is important to get the theology right. According to recent news reports, almost the whole of Britain has become virtually atheistic. A proper understanding of theology justifies and strengthens faith. According to recent British Crime Survey figures 20 to 30 million people are victimised by crime every year in the UK.

Even so, God's laws are written in people's hearts whatever their beliefs. I believe that such people are not spiritually dead (the word 'dead' is not mine, but came from the Holy Spirit in one of my spiritual experiences). Make of this what you, the reader, will. I think it refers to unbelievers in Jesus, or people who cannot or will not see the difference between truth and falsehood or lack Christian empathy. Jesus also used the term in Mt. 8 22, "Follow me, and leave the dead to bury their own dead."

Jews believe in Jehovah as God. Muslims believe in Allah as God, but Jesus taught us about God as Father in heaven. Christians can't believe that Allah is God but they do accept Jehovah. Muslims accept Jehovah but say that the Jews have corrupted Holy Scripture and that Mohammed received an updated version, or the true version. The writer accepts Jehovah in part but rejects much that Jews accept (or used to accept). He rejects Allah and his teaching except as relative to Arabs who haven't changed from medieval times except that they believe in one God and don't drink alcohol, but they are intolerant towards Christianity (as the Jewish priests were to Jesus and his followers). The writer regards individual's thoughts and deeds as more important than their religion, which is more or less forced on them. Thus all Muslims are not misled and all Christians are not inspired. The reality is what counts as true virtue or sin, not appearances. The writer's universal God sees reality and knows the truth. The Father God of Jesus is the Universal God. As Jesus and God are 'one', so we should be 'one' with Jesus. This is the whole purpose of the divine incarnation.

The Supreme Being interacts with the universe constantly. Divine logic (intelligence raised to the nth degree) and natural logic (nature's laws, as we discover by scientific observation) are basically one in principle. If we could think like God, we would see nature in a very different light to the way human minds see it. In fact, logically, if God exists, divinity incarnate, a visible god on earth, is a possibility, because such a being would be a finite presence of the infinite God, or have communion or oneness (a merging of consciousness appropriate to the situation) with God. Not only that but also any number of 'gods', or god-like immortal beings, could coexist. God as in the 'kingdom of God', the Supreme Being, is omnipresent and omnipotent. The universe is big enough for billions of earth-like or life-bearing planets.

One difference between God and a finite being

It is important to understand the difference between God and a finite divine being in present time and place, with an identifiable body, and a single point of view and consciousness like us. God is infinite, omnipresent, has no identifiable body, is spirit, is omniscient and omnipotent, and is not limited to place or time. God can limit his expression to, or in, an incarnate being who would not need infinite knowledge or other infinities.

The way divine incarnation (God expressing himself in a living body) works is by God being in perfect control of everything. There are no 'hypothetical' situations because God controls the environment, and God is the perfect artist and dramatist. We live by God's rules willingly or otherwise. The more we follow our own will in opposition to God the more we dig our own grave. I think Shakespeare was right about the world being a stage, and we are but the players. But this apparently fatalistic 'belief' does not exempt us from doing what we should, or should not, be doing. God is just. Only at the highest level of consciousness would we truly know God, that God and self become one

God is invisible and undetectable. We feel independent and responsible for our own thoughts and actions, but is it God's Spirit working subconsciously through us, or does God send other spirits to possess us when we do evil deeds? We are only conscious of emotions and thoughts which we willingly adopt or resist.

Our world is not divine to us, but neither are we. If we were, our world might become divine also. How we experience the world can change. The presence of divinity causes appropriate reaction. Jesus has given us justifiable hope that our future will be as divine beings in the kingdom of God.

Noah

God said unto Noah. "The end of all flesh (mankind) is come before me; for the earth is filled with violence through them; and behold, I will destroy them with the earth. Make thee an ark..." This should be a warning from the past for violent people who think they are pleasing God with a culture implanted from birth and reinforced throughout their life. They bring their religion, their race and their culture into disrepute. The Lord God said, "Vengeance is mine" (Deuteronomy 32. 35) (Rom12. 19).

From primitive times man has evolved with the idea of tribal identity and survival, especially in warfare. 'What is good for the tribe is good for me and for everyone in the tribe'. Thus, love for the tribe is seen as a virtue. Carried a little further and it becomes love for your neighbour (in the tribe). But tribes compete with each other for territory and food. The more cooperative the tribes become the larger and stronger the group of tribes, forming a nation.

But nations, through population growth, need more land, then we get wars. The cleverest make sophisticated weapons, but their enemies learn from experience and they get similar weapons. Now, at the present time, many nations want the most dangerous weapons. The idea of love or cooperation is far from them. Apart from immediate destruction there is the danger of radiation poisoning.

Christian teaching provides the last step in the development of neighborly love, even to love your enemies. This is clearly too much for some people. It doesn't seem to make sense to them. It evidently makes sense to God, and he wants us to find alternate solutions to our problems and to have faith.

What we choose to believe depends on God, and our characteristic mode of thinking. If you would like to believe in Jesus and a Christian God, but can't, do what many have done before: try it. Behave like a Christian. Don't *impose* your will, but be humble. Give up the obvious vices or you are most likely to fail. They are the outward and visible sign of an inward and spiritual poverty. You might become a saint, and saints don't just believe, they know. This is why Jesus came, to save sinners.

Whatever people believe is what God, in his judgment, causes people to believe. We predispose God to act as he does because of our own predisposition or nature. God is always righteous, which I would call 'divine discipline', the discipline of divine behavior. God takes into account all factors when making a judgment and cannot err or sin. God does not have to think; he knows. Nature proceeds in a logical, intelligible fashion like God.

Chapter 23

Religious experiences UFO's Love You are gods

The universe does display miraculous properties from time to time, and many people, including myself, have had traumatic religious experiences which defy conventional explanation, that satisfy those concerned that God exists, and that he interacts with his creation.

But this would not be admissible scientific confirmation. It is called anecdotal evidence. These days, it would be like saying you have seen a UFO that defied fundamental laws of physics. It tends to damage one's reputation in some people's eyes. If we talk to God it is called prayer, but if God talks to us, that becomes suspect. I would not mention my religious experiences because of the likely negative effect. However, I think that non-Christian people would be more inclined to believe a person with religious experiences than one without.

One experience happened when, one evening outside, I said to an acquaintance, "the Holy Spirit is like a force (not a person)". Immediately, my body turned cold. I knew I had said something very wrong. As soon as I returned to my room I prayed for forgiveness then, as suddenly as I turned cold, I felt warmth flow back into my body, to my great relief. I understood the meaning of the name 'Comforter' given to the Holy Spirit. It bothers me when I hear people say they believe in God and Jesus, but not in the Holy Trinity.

There is another 'religious experience' that I have had from time to time (still very rarely and a long time ago) that might interest people: I have been suddenly aware of 'spiritual beings taking an interest in me when I have written something (religious) unusually perceptive'. One communicates with another, and then others as they gathered quite quickly, and then nothing. I felt joyful to receive this mental evidence that life exists on a spiritual plane. They had no form and I have no idea who they were. They seemed to be like people rather than angels. I never thought of trying to communicate with them. Only one was a strong experience; one or two others were fleeting, but unmistakable. All of my religious experiences happened a long time ago. Also, they didn't speak, they seemed to be telepathic, and they did not seem to have bodies but they had position and local presence. I wasn't asleep at the time, I had just been writing and thinking about Jesus as I now recall.

My most traumatic religious experience happened one evening at dinner in a private hotel where I was boarding permanently. I looked at the meat on my plate, which seemed to represent my father, and I felt compelled to eat it. I resisted the feeling but a clear voice urgently told me to eat it, which I obeyed. Then I felt compelled to find a new father. There were about a dozen people at the table. The man I chose was the proprietor. I said to him, "You are my father". I stood before him in a state of shock and he took me to my room and called a doctor. My mind was affected for about a week, or perhaps it was my eyesight because things (especially people's eyes) looked peculiar and my capability and confidence were affected, otherwise I was reasonably well. I found no difference in myself afterwards. This happened in my very early twenties, about sixty years ago. I know, or feel sure, that the Holy Spirit spoke to me.

If it had any religious meaning it has not made any noticeable difference to me, except possibly much later in my writing or thinking. Now, since I 'solved the riddle of the universe' to my own satisfaction I can see some sense in it. I spoke (much later) to a psychologist who said the experience was unique as far as he knew. My mother died when I was two, and my maternal grandmother brought me up. My father, a kind man, married again to a very nice lady and, later, I lived with them for a short time. My early family history could have some relationship to the event but it felt 'religious' to me, and still does.

Is it a coincidence that I have written and thought (the two are usually simultaneous because I think best when I am writing) about God and the universe? I think not. Is it going to prove a waste of time? Again I think not, or rather I hope not. In other respects I am probably 'normal'. My writing seems to repel religious people but I think it is because they are so used to criticism that they instinctively shudder at anything new. They really have nothing to worry about because I support their main beliefs, just that my way of arriving at them may be different to theirs, and that can be upsetting at a personal level. Many of them take the whole Bible very seriously as literal, but the purpose of all religious teaching is to bring us to a true knowledge of God. Genesis was written for people without any scientific knowledge and today it sounds childish to educated people but, if God is the first cause, that nature obeys God, then man was created by God but not in the way described in Genesis without interpretation.

The scientific explanation for a 'spiritual plane', or different planes of existence would, presumably, be extra dimensions only partially related to our space-time. From Jn 14 2: "In my Father's house there are many mansions…" What other 'natural' explanation can there be? There is no known limit to geometrical constructs of higher dimensions.

UFO's in God's plan

The Bible insists that all man's strivings will end in disaster on the Last day (whatever the cause) and near universal death, then divine intervention will save a remnant. The writer believes that UFO's that have been revealing themselves to us with spasmodic appearances will finally deliver us and provide a New World as the Bible promises.

The writer has fairly recently spoken to three witnesses who are not the kind of people who would lie, one of whom is a relation. In two separate incidents they spoke spontaneously as I was speaking about life on other planets in casual conversation. The accounts confirm what I would expect of very advanced beings with very advanced technology. I have spoken to three of five first hand witnesses and become convinced that UFO's exist. In *two* reports by two people in each report, fifty years apart, a UFO was clearly seen hovering closely above water, but the surface was not disturbed as it would be with a downward force (as from a helicopter). This defies one of Newton's laws of physics. One sighting (mentioned earlier) from an airfield of a UFO that nearly landed and then took off (and was observed on radar) reported a speed of 3,500 miles (5,600 km) per hour.

I think that they travel here in their own contracted time frame, using multi-dimensional technology (no friction and, in some cases, no sound) and I believe their purpose is to save life on earth before our complete self-destruction, doing God's will – as divine beings always do. Divinity is a whole new separate subject. We have to become divine in order to survive. The more capable we are especially in warfare, the more important it is to be morally responsible.

The theory behind the existence of divine beings throughout the universe is that all evolutionary life forms have to pass through the stage where mankind now stands. For the first time in history mankind is capable of destroying itself. Without divine intervention and a miracle of transformation (salvation) it would be likely to happen in the foreseeable future. Scary tales of abduction are unlikely to be true.

UFO's as servants of God

It disturbs scientists that there is no scientific evidence of the existence of UFO's, like there is no scientific evidence of spiritual matters. This might imply that there is an association between the two kinds of events. It is as though it is a deliberate policy of God not to allow the kind of proof that would convince the sceptics. It appears to me that there are people who are not meant to receive faith. Their lack of faith in people's accounts, for instance, without concrete evidence, is counted against them. The words: 'God', 'Jesus' and 'faith' and 'belief' are practically synonymous. Jesus berated people for their lack of belief and stressed the power of faith.

Paranoia

There is plenty of evidence of paranoia and bad judgment in ambitious and ruthless leaders who seem to be determined to arm themselves with the most destructive weapons. There is also the alarming prospect of global warming with unknown consequences and the possibility of even more destructive weapons. Someone asked, "Is deceit justified in war?" The answer came: "Deceit *is* war". We are warned in the Bible that Armageddon or Judgment Day will come upon us unexpectedly.

Seek truth

I believe we should seek truth, and not bend the rules, however well intended, to suit our beliefs or philosophies. Religion is a disturbing and dangerous subject because of the differences in people's deep feelings and beliefs. Nevertheless, it would be in context to suggest that the theory of relativity applies to religion as to physics. Relativity also applies to divine revelation.

Israel

The Jews in Israel have earned the admiration of the world, because of the difficulties of surviving in a hostile region, for their relative forbearance, for their rapid development of industry and commerce giving a good standard of living and for the improvement and productivity of their country. Without goodwill and unity of purpose a society cannot thrive and develop its full potential.

The laws of economics (supply and demand) also tend to discriminate against sectors of the population, but the explanation is usually natural and simple to understand. For instance, large families increase the supply of labor and competition for jobs. The poverty trap and consequent lack of 'seed capital' tends to perpetuate the lack of higher education and training or business opportunities. To him that has shall more be added, but to him that has not, what he has shall be taken from him. This seems to be a general rule. The justification of this principle is motivation, which most people need to elevate their status in life by money, power, knowledge or skill. Progress requires sacrifice, effort, direction and faith or hope.

Love

Love is the outstanding feature in Christian teaching. Selfless love strengthens resolve and motivates people to do the most courageous, unselfish and outstanding things. There is no greater or purer power than love. God sent Jesus to save mankind, and Jesus commanded us to love God and our neighbor. He suffered and died on the Cross for us with perfect faith in God that his sacrifice would not be in vain. God, in righteousness, gave Jesus the power to resurrect his own body and to give eternal life to whom he chose. The power and knowledge invested in Jesus is the power of God, for Jesus dwells in God and God in Jesus.

From St Paul: "If I speak in the tongues of men and of angels but have not love, I am a noisy gong or a clanging cymbal. And if I have prophetic powers, and understand all mysteries and have all knowledge, and if I have all faith so as to remove mountains, but have not love, I am nothing. If I give away all I have, and if I deliver my body to be burned but have not love, I gain nothing...So faith, hope, love abide, these three; but the greatest of these is love" (1 Corinthians 13. 1-13 abridged).

God gave Jesus divine power and knowledge to help people believe. His life, teaching, example, death on the Cross and resurrection justify belief. If people find the Holy Trinity too difficult to understand, it doesn't matter. Just believe in Jesus and what he taught and be saved. If we believe in God but not Jesus, we err. If the writer does not mention churches it is because he believes all men err, no one is divine. If anyone were he would be immortal.

"And Jesus increased in wisdom and in stature, and in favor with God and man" (Luke 2. 52). This is how we should model our own life.

St Paul

St Paul expressed the higher element of Christian ethic in Galatians 5, 17: 'Carnal desire strives against the spirit, and the spirit against the flesh.' The sinless life of Jesus illustrates this better than words. Where the Son is, so also is the Father. Honor the Son and we honor the Father; dishonor the Son and we dishonor the Father. If we do not understand this we do not understand God. If we reject Jesus we are left to the judgment of Nature, which is inevitably death.

People who are desperately poor, especially if they are surrounded by people who are relatively rich, would be more inclined to be militant than the rich or comfortable. But why are they poor? Giving aid is only a temporary solution. Acquiring useful skills and improving education would enhance their status and value to society, more importantly is an improvement in character which improves everything. Christian ethics or standards should improve social relationships. The point is that a happy society has much to lose when violence or war erupts, but an unhappy one has little to lose.

You are gods

Our self-perception is flawed because we are flawed. If we were perfect we would truly know God. As Jesus said, "I and the Father are one (Jn.10.30). He continued his explanation (ref.31-39) "Is it not written in your law, 'I said, you are gods'. If he called them gods to whom the word of God came (and Scripture cannot be broken), do you say of him whom the Father consecrated and sent into the world, 'You are blaspheming,' because I said, 'I am the Son of God'? If I am not doing the works of my Father, then do not believe me, but if I do them, even though you do not believe me, believe the works, that you may know and understand that the Father is in me and I am in the Father." It was this claim that upset the priesthood, because they neither understood nor believed, that led to his crucifixion – which was predestined anyway.

Chapter 24

Addictive drugs Darwinian theory Drama Consciousness Amoeba to Angel About Jesus

Satan

Satan may or may not exist; it makes no difference since there are many worldly causes of temptation. It happens that, as a personification, Satan behaves like an evil and deceitful tempter. Satan does not sin himself, he tempts us to sin (although that is a sin for us). God speaks to him in the Book of Job in the Old Testament, but it is clear that God is in full control. As the arch tempter Satan fills the roll perfectly, particularly when he tells us (intuitively) that we will not be punished for sin or other lies.

Pure evil is causing harm, pain, damage or destruction for the satanic pleasure it gives perhaps because of the feeling of power in otherwise powerless people. It appears as mindless because there is no profit in it, therefore it can be regarded as a mental aberration or sickness. A similar mind-set is defiance or rebellion towards convention, authority, conscience, normal standards of behavior etc. The reward is a building up of the ego, status or sense of power and freedom to do anything one wants. Notorious criminal gangs (congregations of like-minded people) seem to have this philosophy. They rule their domain by fear. Strangely, some people admire them, or some aspect of them, possibly their physical strength and toughness.

The downward path leads to a disaster from which we may not recover but with a new insight to our own condition, and with God's blessing and faith, we may recover. Young people especially are vulnerable to drug addiction because of the feeling that it is not addictive, but this is a mistake. The addiction grows but so does the victim's deterioration in mind and judgment, in the same way as with any sin. Degradation leads to loss of self-respect and loss of good judgment.

Many people think of sin as being a major or important sin. But the truth is likely to be that the minor sins lay the path for the major sins and undermine or corrupt our personality. If we didn't commit minor sins we would not be tempted to commit major sins. To test what is a minor sin apply the golden rule: do not do to others what you would not like done to yourself. Speech and attitude can do more harm than physical violence and be remembered long after afterwards.

Be warned that addictive drugs are the cause of most crime, to get money for evermore drugs. Their cumulative effect is degradation and enslavement where addicts become so desperate that they hurt the people they most love. All addictions are enslaving and have detrimental effects on physical and mental health. They are the result of the body becoming dependent on them. Drugs tend to copy natural substances already available within the body, which do what they are supposed to do at appropriate times, so there should be no need for additional substances.

The tragedy is that young people think that they 'know better' than older 'boring' people and that they can keep control. "Whoever sins is the slave to sin" (John 8. 34). Older drug dealers know instinctively who will likely become the best customers and target them, and then they become drug dealers themselves and become part of the killing chain. Dependence on addictive substances is a way of committing suicide.

Demonic possession would now be called psychotic behavior. We should not be too sure that we always have the only interpretation of illness. The spiritual interpretation is perfect if it works, otherwise a medical approach, along with a wholesome holistic approach, would be the sensible way. Either way would be God's way, but relying solely on a miracle could be interpreted as tempting God, or trying to impose our will on God. We have to help ourselves if we want divine help, by doing everything we can within our ability and opportunity. Relying on faith alone to be cured of a serious illness is like trying to walk on water when you can't swim. God rewards those who make the effort to cure themselves or take medical advice; God works through all things, worldly and spiritual. Then power of faith grows but only in appropriate circumstances.

We should do what God would want us to do within the teaching of Jesus. If it is unchristian it is ungodly. Prayer and faith are best used as support to good medical practice and commonsense. We do not know what causes illness, but my observation is that sometimes a simple solution, like reasonable exercise, good habits and eating a variety of good food prevents serious consequences. Aspire to a healthy mind and a healthy body.

Drama, nature and God's purpose

The intelligent design of nature is evident everywhere. Nature creates order out of chaos because it is designed to do so. Elementary Darwinian theory that random mutation is all that is necessary for evolution seems to me to fall short of requirement. Nature has demonstrated that it is capable of solving problems, especially of survival, improvement and adaptation of species that exceed man's current level of understanding. Nothing happens without a cause. Over millions of years Nature has produced a great variety (many thousands) of life forms, of which man is unique with sophisticated speech and high intelligence, but lacking in wisdom and spiritual insight.

Nature has been described as an experimenter. From a gaseous or molten state in the formation of planets, atoms and molecules are thrown into every possible form of design. This is a method of finding what works and what doesn't. Eventually, primitive organisms, using the same principles, might discover a recurring pattern that speeds up the process of evolution. In other words, an adaptive principle that we haven't discovered. One that we have discovered is death and reproduction, which certainly helps to facilitate reproductive changes. Perhaps there are many more.

Efficiency of conscious thought.

Another new idea is that without conscious thought our brains would be very inefficient. Computers are 'morons that are very fast at adding up', but they have to be programmed to answer specific questions mindlessly, or programmed to moderate their own programs, but they are not conscious. Awareness, from the evolutionary and survival point of view, motivates us to seek power and knowledge of what might be helpful to us.

Awareness is programmed into us from our ancestors but then awareness programs itself using its own logic. Self-awareness is like capitalism while pre-programming is like socialism. One is instantaneously adaptable and personally highly motivated while socialism is not so highly motivated and slower to adapt. A competitive natural environment separates the systems.

Amoeba to Angel

Is it not implicit in higher religion that we should rise above our primitive urges that lead us into sin? That we are somewhere between animal and angel? The body sometimes controlling the mind, instead of the other way round? It seems to fit the theory of evolution, a gradual but dramatic rise from amoeba to angel, with divine help on the last few steps to the otherwise unattainable peak, which is divinity.

About Jesus

If we don't believe in Jesus we should ask ourselves why the disciples, who were ordinary men, continued to believe in Jesus if he wasn't resurrected, and became extraordinary men preaching his message in the face of persecution and death. They were also inspired and empowered by the Holy Spirit. Had Jesus not appeared and talked to them after his death, they would have gone back to their previous livelihood. Also, the social part of his teaching is the basis of civilized behavior, and the religious part its justification.

The logic of salvation is evident: forgive that you may be forgiven; do not judge others lest you be judged likewise; love your neighbor as yourself (so that God may be justified in loving you). The power of Jesus to raise the dead was demonstrated at his own crucifixion and resurrection, which he foretold (Mt 16 21, 22 & 20 17-19; Jn 2 29 & 6 35-40). The reason why he told us to take Holy Communion, symbolically in the sacrament of bread and wine, eating his flesh and drinking his blood, is that we become one body with him. He is not raising a sinner to divinity; he is raising a member of his own mystical body with the consent of the sinner.

It is the sinner's faith in Jesus that justifies his salvation and elevation to divinity, but the power to save is vested in Jesus by the Father. It is also the Father who draws the sinner to Jesus. Remember that we should do our best to predispose God to save us, rejecting evil and self-willed thoughts. In Christianity it is called 'repentance' and 'humility'.

I could never see why sacrifice of animals atoned for sin. I could only equate sacrifice with loss of something valuable to the sinner; in other words, pain or suffering, whatever the cause. I can understand how suffering can expiate sin, at least to the sinner if not his victim. Suffering evokes sympathy in what I would call 'decent' or empathetic people. Intuitively, I think that when suffering has induced a change of heart in the sinner, that would be the time to end the suffering. It would have served its purpose. Sacrifice of animals is a sign of a barbaric culture, and of human sacrifice an evil one.

Unnecessarily inflicted suffering would be cruelty. People who delight in cruelty are without conscience or empathy, and are dangerous psychopaths. Nature, however, enforces destiny apparently indifferently. Can children be taught to be empathetic? Are criminal psychopaths the result of insufficient moral education in childhood? Or are they born hopelessly bad? Or do they represent the extreme in selfish behavior in a society that varies within extremes with most people within the median or average range?

Selfless love is divine. God is love, and light (enlightenment) (1 John 1. 5-10 & 4. 8-14).

The Word that Jesus taught was from the Father. "For I have not spoken on my own authority; the Father who sent me has himself given me commandment what to say and what to speak. And I know that his commandment is eternal life (John 12 49,50)." Who else can make that claim, work miracles, and resurrect his own body?

Heaven to one mind might be hell to another because of he differences in people's opinions and tastes. In the resurrection all will be enlightened, and their bodies will be divine according to God's will. "I tell you, among those born of women none is greater than John; yet he who is least in the kingdom of God is greater than he" (Luke 7 28). Women form about half the population of the earth. What happens in the kingdom of God to the man-woman relationship? We are never told except that there will be no marriage and we shall be 'as angels'. One credible answer is that all will be spiritual, neither male nor female. There will be no death and therefore no children. What does this imply to us at the present time? It is clear that we should live without sin, resisting temptation and following Jesus, and that fleshly desire strives against the spirit. This goes against the present needs of life because without children all human life would disappear. To overcome this problem the institution of marriage in the Bible is accepted as the honorable means of raising a family.

The idea has come to me, in seeking a more rational explanation of the resurrection than miraculous reincarnation, that existing Christian people will become aware of their higher spirituality and become physically, as well as spiritually, over a short period, godlike: fit for the kingdom of God. Our identity lies not in our body but in our soul. The bodies of the dead are of no significance for obvious reasons. While we have no power to change our bodies if we are ill or aged, as a god, or divine being, the body miraculously responds to the divine head. This would be an act of God. Perhaps the body becomes an extension of the mind and can change as Jesus changed from being a spiritual being to an incarnate being to speak to the apostles and then back again. The whole question becomes academic because we cannot imagine how *we* will really be. What concerns us is how we should act in the present to prepare for the future.

It is clear from the Bible that Jesus did not look divine, or like a god. His half-brothers didn't see him as a god. Nobody could have resisted him in his divine body and personality. His power was hidden or he kept it hidden. If it were not so he could not have undertaken the task of the Savior, going to his death, suffering on the Cross, humbly obeying his Father, God, with absolute faith or knowledge. The whole event of Jesus' life and work must have been preordained by God to provide the justification and method for the salvation of the elect.

The immanence of God

On the subject of the immanence of God, I mean the pervasive presence of God the Father throughout the universe including all living things. No one has seen God. The important aspect to us is the way we experience God. Am 'I', or any one of us, God? Obviously not! Do we all, from time to time, do something 'divine'? Quite probably. In this case God is in us and we are inspired but we may not be aware of it. Jesus was aware of the Father's presence in a way that the rest of humanity is not. This is what I call the elevation to the divine estate, and requires sinlessness. The gifts of the divine estate are immortality, God's peace, wisdom, knowledge, divine inspiration and miraculous power. We feel inspired when, though we don't know how to do something, inspiration comes to us and shows us the way; the feeling is rapturous.

It is no good letting God take over your life, in the sense that it is not 'you' doing things. We are designed to exert self-will, but in a way that parallels God's will, we hope. The guides are Jesus and our own intuition. Leadership ability is the ability to make good decisions. Without a strong sense of self-will (intelligently applied) we are likely to fall prey to temptation. God will make his presence known when we are ready.

"If a man love me, he will keep my words; and my Father will love him, and we will come to him and make our abode with him (John 14. 23)."

Chapter 25

Religion a continual challenge Thorny question Hell or Purgatory

We all seem to have an inner need for a challenge and give meaning to our lives. Religion, which has a relevance to everything we do: purpose, rightness, responsibility, quality, duty, integrity etc. can give us both. Some understanding of the complexity of the universe, especially of the aspects that we cannot even hope to understand in detail, helps us to believe in things that we might otherwise have thought impossible.

The most important part of religion is how we express it in the way we live, and how we relate to other people. Are we perfect, as God is perfect? Are we as enlightened, as Jesus would have us be? Do we close our minds to any possible criticisms of our beliefs, such as may be found on the Internet? Are we judgmental and unforgiving as Jesus taught us not to be?" We all have different talents and naturally, a calling to the life we choose. God cares and inspires us as we meet difficulties, if we are pure in heart and humble in acceptance of his will. In my opinion, the biggest flaw in the human personality is conceit, or self-opinion, the classic sin of the Devil.

Many people, especially in the more advanced parts of the world, have moved from total religious belief to unbelief in religion, replaced with belief in science (throwing out the baby with the bath water). And many professing believers refuse to accept self-evident scientific proof, particularly of the age of the earth, because they think it invalidates the Bible. I think that part of their error lies in giving equal, or more, credence to the Old Testament than the New, and interpreting words, such as days as man's days instead of God's days.

Walk to Emmaus

In the walk to Emmaus, Jesus appeared to two of his disciples after his resurrection saying, "O foolish men, and slow of heart to believe all that the prophets have spoken! Was it not necessary that Christ should suffer these things and enter into his glory? And beginning with Moses and all the prophets, he interpreted to them in all the scriptures the things concerning himself" (Luke 24 25-27).

Jesus does not criticize the Old Testament. A sympathetic view is that it is a preparation for the New Testament to be delivered through the prophesied Messiah. The Bible is often described as Judeo-Christian and some churches lean more towards Old Testament values with its emphasis on discipline, form and ritual rather than the more 'free interpretation' of the New. Was the Father of Jesus Christ the same Jehovah God who revealed himself to the Jews? If the answer is positive then the Jewish unbelievers in Jesus have not become enlightened or changed and the answer must be 'wait and see'.

Some modern churches' doctrine is openly based on Jehovah rather than Jesus, whom they denigrate to suit their leaders' level of comprehension. They also have trouble with the doctrine of the Holy Trinity. Where there appears to be a difference a Christian Church should follow Christ as the present expression or Word of God, for the purposes of enlightenment and salvation.

Jesus is famed for his commandment to 'Love your neighbour', but it does appear in the Old Testament: Leviticus 19. 18: "You shall not take vengeance, nor bear any grudge against the children of your people, but you shall love your neighbor as yourself". And again in verse 34: "The stranger who lives as a foreigner with you shall be to you as a native-born among you, and you shall love him as yourself, for you lived as a foreigner in the land of Egypt. I am Jehovah your God." Wisdom and peace are also taught as in Proverbs 17. 9: "He who covers an offence promotes love; but he who repeats a matter separates best friends".

The major doctrinal difference between the Old and New Testaments is the introduction of Jesus as Christ the Savior and Son of God, only through whom is salvation possible.

It is true that God is omnipotent and omniscient, and as our heavenly Father is the source of all things, but his revelations to man is relative to man at any one time, according to his judgment. The purity of Jesus assures us that the Father's revelation to him was the highest level of revelation. Jesus' life, works and resurrection were proof of his divinity. God sent Jesus to be our Savior. If we do not believe Jesus we probably do not deserve to be saved. God cannot sin; he requires justification to save sinners. Only the divine Jesus is that means of justification. Faith in God is natural; faith in a man is not. Faith is the test. In other words, for the purpose of salvation God sent Jesus, not only to teach but also to be the means of salvation – the test of fitness for salvation. What happens to people who see Jesus as the God-sent Savior but reject his divinity, also Unitarians who reject the Holy Trinity doctrine? The writer supposes that God will judge them like everybody else, so the answer depends on God's judgment. If it is God's will Jesus will preach to them after death to fulfil righteousness. Jesus preached that his followers would not be judged.

From John 5. 22-24: "The Father judges no one, but has given all judgment to the Son, that all may honor the Son, even as they honor the Father who sent him. Truly, truly, I say to you, he who hears my word and believes him who sent me, has eternal life; he does not come into judgment, but has passed from death to life."

Many do not believe in the Holy Trinity or that Jesus is divine. This is not as Jesus taught. If I am correct we should look for fault within ourselves, because when we sin we become more corrupted and, where there is no remorse, more ignorant than we already were. We become partially blind or mentally warped, and we can degenerate with antisocial ideas and deeds. Sinful desire indulged cannot be controlled, it grows. We have to put sinful thoughts out of our mind to prevent the weed of sin taking root.

We are all sinners, so why do some people believe and some do not? The simple answer is that this is God's judgment on the unbelievers, but which factor, or factors, led to this judgment?

Change and become like little children

Two factors spring to mind: (1) "I tell you the truth, unless you change and become like little children, you will never enter the kingdom of heaven. Therefore, whoever humbles himself like this little child is the greatest in the kingdom of heaven." (Mat.18 3,4). (2) From Mat. 7. 13-23: "Enter through the narrow gate. For wide is the gate and broad is the road that leads to destruction, and many enter through it. But small is the gate and narrow the road that leads to life, and only a few find it.

"Love your neighbour as yourself." The writer would like to express this in modern language to make it, perhaps, more understandable: Empathize with all people, friends and enemies; see them as they see themselves and gain the gift of understanding their point of view, and why they think as they do. Then act appropriately in a civilized way. Follow the teaching and example of Jesus.

(Mt 7 15) "Beware of false prophets." They come to you in sheep's clothing, but inwardly they are ferocious wolves. By their fruit you will recognize them. Do people pick grapes from thornbushes, or figs from thistles? Likewise every good tree bears good fruit, but a bad tree bears bad fruit…Not everyone who says to me 'Lord, Lord' will enter the kingdom of heaven, but only he who does the will of my Father who is in heaven."

I think that this last statement requires some explanation, since no sinner can ever do the will of the Father perfectly, at least in the absolute sense. Faith, love, repentance, humility and the readiness to forgive, in the name of Jesus, are the hallmarks of a Christian personality. We must surely be judged on this basis, remembering the emphasis on faith in Jesus being the spring from which the living water of Christianity flows. Furthermore, all must be considered relatively to the circumstances of the sinner.

From he to whom much has been given, much is expected. Also, Jesus said that believers "shall not come into condemnation, but passed from death unto life" (Jn 5. 24). I think the important aspect of a sinner's personality (it applies to all) is true humility. That would be the difference between the saved and the unsaved. Believing Jesus is believing in God who sent Jesus to teach us the way to God.

Thorny question

On the thorny question of personal wealth, I think that the important question is how it is managed. Is it used for self-indulgence or is it used responsibly in public service helping people and the economy? The concentration of wealth and capital investment is necessary in the modern economy to raise the standard of living of all the people. The subject of economics, like politics, is full of traps to mislead. The truth is frequently the opposite of what many people expect or want to hear. Essentially, good people create a good society; bad people spoil it.

Which religions teach us to hate sin but love our enemies? Even primitive tribes love and support each other for war or defence against their enemies – who are probably not much different from themselves. Only Christianity teaches, 'love your enemies'. But it doesn't say give them weapons or help them to be anti-social; it means enlighten them and help them to be better and more loving people.

Salvation to the kingdom of God is a raising from human estate to divine estate. It is offered to everyone, regardless of rank or condition. It does not take great learning to know the difference between good and bad deeds in the Christian ethic. What is required is the humility to admit to ourselves, that what we might previously have thought was right, was actually wrong, and to follow Jesus.

Just how we actually do this is intuitively guided, but we should be sensitive to other people's feelings, lest we do more harm than good. The presence of Christians in the community should be an inspiration to others, not an offence. The basis of good manners is Christian love. God is love, and God is light. Jesus is the Light of the world. Enlightenment can only come from God.

Hell or Purgatory

On the subject of hell, once again seeking a materialistic explanation, nature appears to treat life as a continuum within a family; the virtues and vices of the ancestors are passed on to the descendants. How could this be just? Presumably the virtues offset the vices. In human judgment we feel like individuals, but in fact our knowledge of self is limited to our personal memory and experience. Our body inherits genetic characteristics, but what of the soul? Is it immortal? If it is, then our association of the sense of self with the body (which changes and dies) is temporary.

Many people do not believe in hell, but a kind of hell exists all around us, especially in the more unhappy parts of the world. A better name for this would be purgatory where we can rise by pleasing God, by resisting temptation and making the world a better place; or fall further by being uncaring, selfish and self-indulgent, exploiting people whenever the opportunity occurs. People have been living in this kind of purgatory since the fall of Adam and Eve.

If this is purgatory, what is the biblical hell? It is described as hell-fire, and is clearly meant to inspire fear. If death is believed to be the end of life, meaning destruction, it holds little fear for evil-minded people. Everybody dies anyway. Logically, to them, it should make no difference whether they commit small crimes or heinous ones. If they feel even a little doubt that oblivion may not be synonymous with death, it may prevent them from doing more evil deeds. Most of us take more notice of our feelings than our thoughts.

The rulers of this world, where they make life more hellish, are 'possessed by demons' that derive their pleasure and satanic sense of power by making their subjects suffer. Satan is the personification of evil.

In conclusion, I feel that the world has a real need of an update in Christian doctrine to bring it in line with contemporary science as I have described it. Instead of being in opposition to science (and losing the battle) become more scientific than the scientists who have been (unintentionally) displacing God by describing a godless universe.

True Christians don't need miracles to believe. Nature itself is a magnificent miracle of God that all would see if their eyes were opened. Jesus is real, people seem to have forgotten that. The world is getting worse and the last days may be closer than we think. Christianity has all the answers to every human problem. God, through Jesus, has made an offer that everybody should accept, but many are blind because of sin. The answer to the problem is clear: turn away from sin and seek God through Jesus who is one with God.

Chapter 26

Solving problems in economics Excessive created money and insufficient real currency

While the subject matter so far has been about cosmology and theology the discussion of which is to enlighten people and improve life both in the present and the future, there are immediately pressing needs. It is a Christian service to help people. In the writer's opinion the best way is to impart knowledge or understanding. Economics might not seem to be a proper subject for Christianity, but poverty exists mainly because of bad economic policy, and poverty is a Christian's problem. In fact, most things concern Christianity from how we do things and how we treat people, to preventing world wars. The good things we *want* are pretty universal, but *how* to get them is the most important area of disagreement in politics.

Share-markets in 2009 worldwide crashed to less than half their previous value, some much more. Banks and lending institutions were failing because they could not meet their obligations and there was evidence of a severe depression looming. Governments have had to inject hundreds of billions of dollars to prevent widespread collapse of the whole banking system, with its inevitable consequence of widespread economic disaster.

Much of the world in 2010 is recovering slowly from the recent depression. The writer would like to contribute his ideas to just the basics of recovery. Very broadly, the money supply (real currency) forms the financial base support for the 'created' money supply. An increasing total money supply is inflationary, causing increased borrowing to take advantage of the expected ever-rising prices. The wave of prosperity reaches its natural peak based on high valuations of property, general prices and shares. Financial commitments are made based on the artificially high values and these commitments have to be met.

Suddenly, one of the over-leveraged borrowers loses the confidence of his lenders and he can't repay the loans in the agreed time. His overpriced assets prove hard to sell as word spreads around. No one wants to be caught with bad debts as that can be disastrous, and no one wants to buy in a falling market until bargains based on historical values appear. So the wave is followed by its trough. Money can't be borrowed without collateral security and that has already fallen in value. Lenders want to be repaid because their loans are now at risk.

The total money supply (currency plus asset-backed loan money) in a depression has considerably decreased. Created loan money cannot be reclaimed because the assets have shrunk in value. When loan money exceeds currency by a factor of five, or ten, or thirty-five in one major firm the market becomes unstable and likely to crash. The reduction of value in the loan collateral security makes it impossible to repay the loan. Currency, on the other hand, increases in value because prices have fallen, but the volume of currency is small in relation to created loan money – in badly managed economies.

The main trouble is that this process hurts the poorer people through massive loss of employment with widespread unhappy consequences making the situation worse. The situation will eventually self-correct with forced reduction of prices and wages (with a constant real money supply that increases in value or purchasing power as prices and wages decrease). The more resistance there is to this unpopular process the longer and deeper will be the trough. The quick remedy is for governments to inject more currency (or created loan money guaranteed by the government) into the economy.

It must be a fair comment to say that, apart from economists whose business it is to take note of long runs of apparent prosperity fuelled by created loan money, and the increasing instability of the system, most people are unaware of the true state of the economy, especially when the economy appears to be in good condition. Everybody knows when the economy is in bad condition. Only governments have the power to regulate the creation of currency (cash) and loan money, meaning bank cheques without holding cash in reserve to the same value. Also, democratic governments have to be careful not to make themselves unpopular and voted out of office. Just printing more money raises prices and wages more or less equally, called inflation, which also distorts the economy and makes savers poorer unless they invest in equities that increase in monetary value with inflation, and borrowers richer whose loans are repaid eventually with devalued money. When people borrow heavily it usually means that they are fairly certain that inflation will increase prices and wages. If deflation were expected they would be keen to lend money.

The point of this discussion is to show that the currency (money) supply governs the basic or average wage-price level. In bad times the currency supply can be used to stimulate the economy, but in over-stimulated times it can be withdrawn to prevent devaluation of the currency (another way of describing inflation). Then there is the fear that it could start another depression. Economists think that a controlled inflation rate of one or two percent per annum would be good for the economy.

Inflation (of prices) also has bad effects, destroying confidence in currency apart from causing high interest rates. A wise government has to know when to apply the brakes to the economy and when to press the accelerator. Only the government can increase or decrease the money supply at will by printing money or issuing credit (backed by the power to print currency if needed), but it must be done responsibly and thoughtfully.

The hard part is withdrawing excess currency to keep the overall level of inflation within acceptable limits. Withdrawing currency reduces taxable income to the government, but also reduces (or restricts increasing) costs. It also reduces interest on borrowing. Money does not make wealth but it tends to redistribute wealth, especially to those who lend, borrow or invest money wisely. The value of money is in inverse proportion to prices. Changing the amount of currency in circulation also changes its value. Higher prices mean that currency has a lower value.

Money creation through the banking system is described in the Wikipedia website. Search 'Money Supply' and 'Fractional Reserve banking'. It makes interesting reading. Banks are allowed to create money (loans) with the government's blessing based on a multiple of the amount of capital (deposits) available, which is assumed to be based on 20% reserve. This means the bank can issue credit of (a variable rate set by the government) $100 for every $20 of reserves. Naturally there are many views on this issue. The actual figure for the total credit issued by banks (indirectly) is probably $200 or ten times the net asset of money held or deposited. This is because other banks receive this extra money in the business process that they legally use.

Taking a generous view, the government is using the banking system as a retailer of loans (which is likely to be a lot more efficient and satisfactory than the government itself) receiving its share of the profits from interest and tax on the income generated. The system is practised worldwide. It seems that the government has to keep a close eye on the economy to prevent unsustainable booms and busts that can be disastrous to everybody. Democratic governments change frequently, which tends to make them short sighted. There needs to be controlling body that thinks far-sightedly. Reckless management of finances on a very large scale is irresponsible, even criminal.

The natural order should create its own market, matching prices of products with wages so there is a natural balance. If there is more than minimal unemployment it means that the wage price level is too high in relation to the total money supply. Either reduce wages and prices or increase the money supply. True wealth depends on efficiency of production. If production efficiency increased enough, workers could increase wages and leisure time.

Placing restrictions on buyers or sellers distorts the system. A distorted system makes some people better paid, but at the expense of others who become poorer or unemployed. A distorted system has to be subject to some coercion to sustain it. A perfectly free and natural system (a perfect market) would smooth out the distortions until full employment and trade is obtained. Sellers would maximize net profit (profit per item times volume of turnover) with free competition; workers would maximize wages until free competition from unemployed workers lowered wages. Easy to say but hard to do! Competition feels bad for the individual or group but good for everybody. We all benefit from healthy competition, because it maximizes true wealth that can then be redistributed (to a degree) through taxation for social benefits.

It should be noted that prosperity depends on efficiency. Efficiency depends on adequate capital investment for improvements and economy of scale; in other words mass production and a large market. There is a very big difference in the individual manufacturing cost of components or finished products when the number required is increased substantially. It has been estimated that the manufacturing cost is decreased by 20% every time the number of components is doubled. Improvements in methods, such as computer-aided machine tools, will have changed this rule-of-thumb in some types of manufacture. It is likely that the percentage of people on the 'factory floor' will decrease and that service industries (like distribution, entertainment, medical etc) will increase, or more sophisticated and expensive products will become available. Training people to become more employable would help.

The recent multiplication of small nations, while great for national pride, or cultural comfort, may be bad for the nation's wealth. They must learn to be cooperative with trading partners to overcome their economic disadvantages. The USA has prospered because of the unification and free market between its many states. Countries that do not have access to large markets are at a (manufacturing) serious disadvantage. Farm production can be substantially increased with economy of scale and availability of capital. Political stability and a high standard of law and order are significant factors in a well-organized society. The need of a large police and military force keeps the country poor and poverty increases corruption throughout the establishment.

As well as having control over the money supply the government should also put a sensible limit to the degree of leverage in businesses where managers are not speculating with their own, but someone else's money. When the speculation is a success the manager gains, but when it results in a great loss he doesn't lose. In the meantime he has usually made a massive fortune for himself from generous bonuses in the good years. In June 2010 it was noted in America that some banks were 'already doing it again', over-leveraging to make bigger but riskier profits (riskier for everybody). "Where are the regulators?"

Banks do not allow their business borrowers to over-leverage. Over-leveraging is like a licence to print money, but at high risk. Governments can do the same (also causing inflation) at very low cost and no risk because they *can* print money.

A little bit of inflation can stimulate a chronically depressed economy. Nearly all the American states are reported to be seriously short of money, why not lend to them with printed money, at a sensible rate of interest, until the economy improves and unemployment drops? When inflation appears that is the time when the government should put its foot on the financial brakes. Prices need to meet the market to regain lost business, or buyers need more money.

The following quotation should give the reader some perspective about the importance of economics:

"Next to war, unemployment has been the most widespread, the most insidious, and the most corroding malady of our generation. It is the specific social disease of Western civilisation in our time" (The Times, 23 Jan. 1943). "Without the great depression between the two world wars [1914-18 and 1939-45] Hitler would never have come to power, the Second World War would never have happened, and it is extremely unlikely that the Soviet system would have been regarded as a serious rival and alternative to world capitalism." Conclusions and quotation from 'Age of Extremes' p. 86, by Eric Hobsbawm). Incidentally, it has been said that the American government of the day made the depression worse by its economic policy than it needed to be.

Quotation on capitalism by Sir Winston Churchill: (1874 – 1965)

"Some see private enterprise as a predatory target to be shot, others as a cow to be milked, but few are those who see it as a sturdy horse pulling the wagon. The inherent vice of capitalism is the unequal sharing of blessings. The inherent virtue of Socialism is the equal sharing of miseries."

Subsequent problems have been how, exactly, is the money to be injected where it will do the most good. The banks have mostly been saved and seem to be doing well, but poor business turnover in the USA has caused high unemployment. This is the area that presently needs help (October 2009). We don't seem to know if or when the injected currency will flow down to the general population and stimulate spending, so perhaps some more money needs to be put into the hands of the end purchasers – and recovered when business sentiment and employment improves, as it eventually must, hopefully sooner rather than later. Perhaps the planned national medical services expansion would be a good start? Building more hospitals while training more doctors and nurses would release more money.

This depression happened because, in an atmosphere of rising prices (inflation), it was usually sound business practice to borrow as much capital as possible at an economic cost. The risk is that if too much is borrowed it can become a disaster; if too little then you fall behind your competitors. Inflation is the enemy of savers unless they invest in equities that should keep abreast of inflation (with the risk of deflation). Excessive borrowing (gambling, or speculating, on continuing inflation) in very large firms affects everybody worldwide, except China which is developing its own economy and infrastructure following a policy of low wages and low prices.

It is possible that widespread inflation, an increasingly modern problem, has caused bigger waves and troughs, certainly bigger waves, than ever before. Excessive house purchase finance in the USA, facilitated by perceived government guarantee enabling 'derivatives' to be used for capital raising, was probably inflationary (equivalent to printing currency) and the first cause of the breakdown in the financial world. If this theory is correct, the government could reintroduce the policy, but on a smaller scale, as a quick remedy to the unemployment situation.

The performance of a business is rated by the profit it can make relative to capital used. Borrowing to increase business capital gives more leverage or power to increase profit on original capital. The negatives are the cost of borrowing and the risk of failure to repay what is borrowed or pay the interest. When banks see the net tangible assets of the borrower shrinking (as in a depression) they demand a reduction of the principal sum borrowed. The urgent sale of assets depresses the market further. Shrewd and long-term wealthy investors eventually buy those same assets, when they estimate the price will not go any lower and the market will recover. The not so wealthy cannot risk further loss.

The economy at the street level usually starts recovery about 6 months after the financial market recovers – according to history, but it seems to be lagging behind at present. With about ten per cent unemployment (2010) wages need to reduce or more currency is needed to raise consumption at present prices. Another way is to devalue the currency and follow the example of China. Then imports would cost more and exports would earn less. This would be unpopular but it would achieve results. Undervalued currency means people have to work harder but the country sells more; overvalued means the opposite.

Overvalued means overseas travel is cheaper and so are imports which is good for prices to the consumer but bad for domestic exporters.

The present situation is so much more complicated than in previous depressions because of a major change in industrialization of China and India, and of the employment of relatively cheap labor in those countries. Pressure is being put on China to revalue its currency upwards, which would make its costs and prices higher to foreign countries and imported goods cheaper. China is already a major importer of mined products as well as luxury goods. Like the USA it is a large country with economic advantages and good workers.

Also, recently, the rapid rise in the price of basic foods was threatening, or causing starvation of the poor in many parts of the world. The recent increase in the price of oil and the subsequent need to find alternative sources of fuel, has taken agricultural land away from food production for fuel production. It has also reduced disposable income in oil-importing countries. If governments want to create more money at the grass roots level they could reduce the tax on fuel.

Unemployment

"Most economic schools of thought agree that the cause of involuntary unemployment is that wages are above the market clearing rate". (Introductory quote from Wikipedia-Internet). This is followed by a bewildering set of individual theories. Employment restrictions are often made for well-intentioned reasons but at a cost.

Philosophically, capitalism is a 'natural' system where people are free to use their own initiative to take part in the economic system of supply and demand, whereas communism is an organized system controlled by the state. Socialism has the same basis but offers compensation to previous owners whereas communism doesn't. Communism promises equality of income, but Russia soon found that didn't work, and remuneration according to perceived value became the rule. Efficiency of production became so poor (and miliary expenses so high) that Russia finally abandoned communism. Efficiency of production (of tanks) during the invasion of Russia by Hitler's Germany (1939-1945 war) was brilliant, but that was because everyone was highly motivated.

Private enterprise, otherwise known as capitalism as its name implies, thrives on capital because that is usually the essential element in setting up a business for profit. Capital makes more capital if it is invested or used wisely. It is the savers that provide or acquire capital. The more capital that is available relative to the demand (or need) the lower the interest rate, and vice versa.

The natural order (tending to extremes of income) is ameliorated in rich countries by distribution of wealth by taxation. High taxes on income lead to high charges for services rendered so there is a limit to the efficiency of taxation. To a degree, high-income people are partial tax gatherers for the government. It has been estimated that the maximum level of taxation for maximum return to the government is 39%, but this may not include the variety of other additional taxes. The lower the tax rate the more competition and efficiency and the higher the national income, but the whole question has to be understood taking all aspects, including health, happiness and leisure into account.

Government tax and labor laws have to consider unemployment, efficiency, work conditions, leisure and fair competition in all aspects of the market. The problem with paying high wages to a privileged section of the total work force means correspondingly low wages (or reduced purchasing power) to the underprivileged. Preventing the sacking of unwanted workers by law prevents the business from hiring new workers at a more economical wage absorbing the unemployed. This is a problem everywhere and is more of a social problem than an economic one.

The threatened workers would be horrified after all their efforts to uplift their standard of living. Schemes to help the poor should make available education or training to make them more valuable as employees or self-employed workers. Unemployed workers have to live so without welfare from the government they have to beg or steal to survive. In other words those in work have to support those out of work. True freedom of movement of workers has benefits to many, but may reduce wages to a few. Overall there should be a great benefit both to workers and the economy.

The economic role of governments should be to establish rules of behavior (law and order) that allows the maximum production with the minimum of effort; and as fair a distribution of that production as is practicable with regard to other important factors, such as reward for effort, without which the system stalls with dire results. I would go further and say that a responsible government would establish working conditions and communication channels in every business that would promote goodwill and employee identification with the employer. The two things they cannot rightly set are wages and prices, which should find their own level with honest competition.

Employers are not hurt by regulations as long as their competitors have to keep the same rules. Only a government can do this. The way to increase true wealth is by greater efficiency. Money is like the good oil that makes the economic engine run smoothly, but the bad oil makes it run badly. A truly happy society is one where everybody feels empathy with everyone else. The end result should be prosperous economy that can afford a good welfare program. The alternative is a poorer economy that can't, no matter how good the intentions.

Professor Robert Wade (London School of Economics) said the top 1% of income-earners increased their share of total US income from around 16% to a peak of 22% in the decade up to the 1929 sharemarket crash. This plunged and bottomed at 9% in 1977. By 2006 their share rose to 22% again. The writer does not know if these figures represent income before or after tax. Lower income worker's real [inflation adjusted] incomes barely changed in the quarter century to 2005 (NZ Herald Oct 28 2009). He also said that rising prices encouraged people to borrow heavily against their houses, and fuel a US trading deficit that by 2006 was bigger than the entire national income of India.

Chapter 27

Ratio of borrowing to capital

The problem of persistent high unemployment (as reported in the USA (2009-2010) is that prices are too high in relation to the supply of money, or the supply of money is too low in relation to prices. Freezing wages (costs) and prices (controlled by wages) and injecting money (currency or loan) will increase consumption indefinitely, or until prices rise due to shortage of commodities.

A few years ago in New Zealand prices and wages were frozen by law, which was very unpopular, but unemployment disappeared and there was a shortage of workers, so the system has been proved to work. Whether it is sustainable or not over a long period has yet to be proved. Shortly afterwards, when import, overseas finance and wage/price controls were lifted, New Zealand had the most expansionary sharemarket in its history. There was even talk that New Zealand, because its sharemarket is the first one to open daily due to its closeness to the International Dateline, could become the world leader.

Due to an aging population the need for pension funds has expanded the capital base. This may well transform sharemarkets because over time, logically, this will be reflected in higher shareprices and lower returns at least until their outgoings match their income. Once again we have to realize that, in economic terms, money will not make bread. The answer to the problem of more consumers and fewer producers is efficiency of production – or better health and strength in old age.

Booms and Depressions

It is notable that Canadian banks have come through the recent depression in good shape. Unlike the American banks they were conservative in their lending, though making less profit. Firms that borrow excessively do extremely well in boom times but go bankrupt in bad times. In fact, it is excessive borrowing that causes booms that inevitably cause depressions. and it is inflation that causes excessive borrowing. It is unfortunate that the effects are so widespread. The solution to the problem is obvious: the ratio of borrowing to capital must not be allowed to be excessive.

It is one of the faults of so-called democracy that individuals cannot be held responsible. A number of professional people warned that high-risk and low deposit mortgages were being accepted on a big scale in 2006, but it wasn't enough to convince the people who had the power to impose the necessary regulations to curb the dangerous practice. The government allowed political considerations to override economic stability. The intention was good but the effect was bad.

About democracy, if the will of the majority is good the result will be a good government, but if the will of the people is 'bad' (because it is impractical or has serious flaws) the result will be what one might expect. It is said that a gifted benign dictator (see the part about Ataturk, the first president of Turkey) would make the best government.

There is a major flaw in simple democracy, the voters are not experts and they tend to elect unsuitable people, but we have to agree that it is generally better than a dictatorship which gets out of control and dangerous, and can't be displaced except by bloody revolution. Dictators and their counterparts, one party states, seem to more keen on armies and armaments for making war and staying in power than getting on with making life better for the common people. They use false fears, propaganda and national pride to boost their popularity, while at the same time they use force to silence or eliminate their critics.

National unity is valuable but if the leadership is unwise it can result in misery and disaster. Revolutionary leaders promising communism for the people's good, setting themselves up as a savior of the people, when they find their plan has failed they don't abdicate. They keep their position and power ruthlessly destroying any opposition. Their saintliness can no longer be supported; they have 'feet of clay'. Hard-working Chinese people have built China's industry and commerce. And now the communist government is probably misleading them again by applying their hardline philosophy to other nations instead of being cooperative. It is creating hostility with possibly serious consequences. China may become a great military power, but they will not be admired.

True (and legal) democracy requires the elected people to represent *all* the people in their constituency, not just the views of their supporters. Unfortunately such high mindedness is rare or impossible to fully implement. Also, they have to take into account realities, or hard facts. They should keep in mind that a prosperous economy is necessary to support welfare programs. A prosperous economy (supporting a tax base) needs to be given priority to enable other sectors like education, welfare, pensions or defence (which seems to be out of proportion to the others in the U.S.) to be possible without going into serious debt. Such debt can only be paid off by inflation; lenders get paid back the principal, or receive interest, in devalued currency.

Borrowing from overseas has the same inflationary effect as borrowing from the government, but can be used for repaying foreign debt. It also imposes a discipline of repayment, plus interest, also in foreign currency. A home based lender (the government) should be better than borrowing overseas. As long as suitable interest is charged on loans it should not be inflationary. Saving money helps the economy if it is used in business to increase production or service.

"Now, unlike the Great Depression, central banks and finance ministries know it is better to run deficits and print money than to suffer massive losses of output and jobs…. Given the immensity of the crisis, a Congress-approved bailout may be just a short-term fix. But a short-term fix is better than no fix. If nothing else, it would signal to the world that - unlike in 1930 - the U.S. is doing what it can to avoid financial calamity and sidestep Depression 2.0." ('The End of prosperity', by Niall Ferguson, Time magazine, Oct, 13. 2008).

Another highlighted quote from the same article: "The real cause of the [1930] Depression wasn't the stock-market crash but a contraction of credit due to an epidemic of bank failures." (Milton Friedman and Anna J Schwartz: 'A Monetary History of the United States', 1867-1960) "As Friedman and Schwartz saw it, the Fed could have mitigated the crisis by cutting rates, making loans and buying bonds. It made a bad situation worse by reducing its credit to the banking system." They made a depression into 'the great depression'.

Arguments that there is not enough work for people are usually not true. There is a lot of work (goods and services) needing to be done almost everywhere; what is missing is enough money to pay for the work, or enough qualified workers willing or able to work for the money offered. Most governments would like to spend much more money than is available to improve infrastructure, healthcare etc. as would business owners to improve their businesses, and homeowners to improve their homes. It is the business of economic management to provide the means.

Business works best when it is motivated by profit. China's spectacular rise began when the formerly restrictive government gave the capitalist horse its freedom. It can be reigned in by taxation, but every increase slows down the horse. There is an economically indicated level at which any further increase hobbles the horse so much that the economy suffers with a loss to the government.

Monopolies

Money circulates, so when business becomes less lively there is less circulation. When money changes hands there is usually some tax collected. Horses run fastest when they are racing against other horses. Competition keeps business competitive and efficient. Monopolies kill competition. Governments and city authorities are monopolies, and (worldwide) they are notoriously inefficient and frustrating. Monopolies are plentiful today wherever they can thrive, causing a large gap between the lower paid workers and the higher paid. In effect, the higher paid are raising prices, usually affecting everybody. Business monopolies can be ruthless if not investigated and controlled. Governments seem to very aware of this and should take swift action. The main cause of unemployment and injustice of wage differences can be traced to monopolies in the labor market. The end result can disastrous for everybody. Monopolies in business for goods or services can result in extortion.

It is in the public interest to give value for money. It is in the private interest to maximize the profit and minimize the service. Only competition provides the incentive to provide a better service at a lower cost. Lack of fair competition often results in extortion, meaning very high prices in relation to cost of providing a service or product.

Higher than average wages, especially in bottleneck (strategic) industries may lower the real income of other workers who are the consumers and also contribute to the level of unemployment. If it were not for increasing unemployment and higher prices wages could be raised everywhere by law. Inflation destroys any gains artificially raised wages (everywhere) might make. Higher than average wages in some sectors is against the interests of the lower paid sectors. If all sectors were highly paid prices would rise in proportion. Unless more currency were released unemployment would increase. The advantage of higher wages can only exist if they rise relatively to other income earners.

In times of high unemployment, because of a smaller workforce the workers would have to support the unemployed and there would be less goods and services. Everybody would be poorer except the unemployed that would be paid for not working. Money is only a convenience of trade and does not produce tangible wealth like real things. Self-employed workers on contract are usually more efficient than wage earners for obvious reasons and can cost less and earn more through higher productivity.

The real problem of wage protection arises because of the differences in incomes. People with specialist skills earn more because employers find them to be more productive, meaning they make or save more money for the employer. In this case they are not taking any money away from the lower paid worker because their efficiency (as money producers) is comparatively greater – in theory. In practice bold, well-informed business gamblers, for instance, can make huge amounts of money unethically because they don't seem to have to pay it back when they lose.

Excessive protection of workers jobs and incomes (by the government) at an uneconomic level, as in S. Africa in recent years has produced up to fifty per cent unemployment and a very high crime rate (information by a late resident). No amount of goodwill can change economic facts. A boom in the mining industry from demand by China and India has been helping S. Africa and other mineral exporters.

Competition is not a part of Christian teaching, but the world is not composed of saints or dedicated altruists. The economic system (means of survival) has to adapt to current society's needs. Governments have the power to change unsatisfactory conditions, but they need to be composed of people with an unbiased appraisal of the truth, with a desire to improve the conditions of life for all people – if they were allowed by the very people they seek to help. There lies the root of the problem: the people. Government representatives have to listen to the voice of the people who elected them or they'd lose their jobs.

Compounding interest

Let us imagine a thrifty bank (or anyone) lending their working capital at a net profit of 5% per year. They reinvest their capital plus the 5% profit on the same basis each year for 10 years. Their capital would have increased from every $100 to $163 (compound interest) in the tenth year. Multiply this by the millions of people and businesses that do this and the result is a vast increase in money supply and purchasing power – or would if prices and wages had not risen proportionally. Notice the absence of increase of real currency, which is the backbone or guarantee of bank or loan money.

Allowing banks to create loan money without paying the government (ie the nation) for the privilege is usurping the government's privilege to issue currency, which ultimately governs the level of wages/prices. The responsibility to save the economy falls on the government when things go wrong, so why don't they make the profit, or some of the profit, to help pay taxes?

At this time of writing, October 2010, in the USA people are 'frustrated' and 'anxious' at the slow recovery of the economy at the working class level.

There is one promising piece of news from 'Time' magazine (Oct. 09) on the Global Warming front: Takashi Yabe, a Japanese scientist with unusual vision, is researching a clean alternative to fossil fuels. It is really two subjects: the magnesium injection cycle (MAGIC) and obtaining power from the sun. The MAGIC engine mixes magnesium granules and water to create heat energy and hydrogen that can power a car. Magnesium batteries are seven times more powerful than lithium-ion batteries (which are also expensive), but magnesium requires a lot of heat at temperatures up to 4,000 C, to refine it and is very expensive. As Takashi (a specialist in heat transference) says: 'Sunlight is free.' A lot of the cost calculations can include the cost of paying for capital (interest), which can vary from zero to an extravagant amount. Let us hope that scientists will make a breakthrough to solve the world's energy problem.

Scientists are making new discoveries that will help solve economic problems but they are presently too expensive to be practical. That is another problem to solve in this troubled world.

An international language

One would think that good communication between all the peoples of the world would be desirable. English, 'the language of trade' has become the most used in international commerce. Apart from its antiquated spelling it serves the purpose well. English spelling is two hundred years out of date because it can't keep up with changes in spoken language. The verb 'to be' also needs updating. There is another good reason to update English spelling apart from making it easier to learn as a second language, a substantial amount of time is wasted in schools just learning to spell. A large part of the English-speaking population is afraid to put pen to paper because they are likely to make spelling mistakes. The English and American dictionaries must be the most widely referenced books in the English speaking world. English is a lot more efficient than many languages and (I was surprised to hear) easier to learn. It also has a very large vocabulary with many shades of meaning, which makes communication efficient, and a massive amount of literature. It has borrowed from and evolved from other languages so it does not have the possible stigma of being an artificial or created one. It lends itself to poetical and dramatic expression. It is continuously changing and growing.

Why people are so obsessed with the importance of keeping their own language when they could communicate with other people much better with a common language gaining a better understanding (and avoiding misunderstandings) seems irrational to me. Why make life unnecessarily difficult? Much of everybody's past culture or behavior would be better forgotten anyway.

A language is just a code for identifying things and meanings, but the sounds probably reflect the manner of the people. Perhaps they take a particular pride in their language. On the negative side it divides, alienates, confuses and separates people who might otherwise be friends. The tower of Babel is a curse. "Its name was called Babel, because there the Lord confused the language of all the earth…." (Genesis 11 9). Minorities try to keep their language alive along with their cultural traditions, which identify them but, in some ways, arouse hostility of others. Inability to communicate arouses suspicion and causes misinterpretation. We learn from, and about, each other. From openness comes light. Evil hates light. God *is* light.

Repetitive Staff for Keyboard or Piano.

The present archaic system of writing piano music on two sets of five lines evolved from the previous system of eleven lines with middle C at the center. After some experimentation to further improve the system I produced the following simplified idea:

Two double sets of staffs (or staves) similar to the present style, the upper pair for the right hand and the lower for the left, formed of two sets of three lines each. There is a total of twelve lines. I have called each set of three lines a triplet. The bottom line always represents C, the middle line E and the top line G. The notes of the 8-tone scale in the key of C, all white keys, are (as now) CDEFGABC. The notes C, E and G make the sounds of Doh, Mi, Soh, which harmonizc as first, third and fifth harmonics:

Bottom line C, the space above line C is D, followed by line E, then by space F, then line G, then by a 50% wider space to accommodate A and B (A in the space just above line G and B in the same space just below line C) making a straight ascending run of notes.

Staffs would be divided into bars to show the timing as they are now. The range of notes on the set of 4 staffs (top 2 for the right hand and bottom 2 for the left) would be a little wider than the present system which uses a lot of leger notes (above and below the staff) which can be confusing to novices, whereas the triple line system is easily identified; the code C, E, G is repeated ad infinitum. The simple pattern makes notes and chords easier to recognize than the present five-line system. Also the five lines in the right hand (EGBDF) are different to the left (GBDFA) which further confuses learners.

The method of notation could easily be changed to linear or graphic representation. A thicker line whose length represents the duration of the note would be much easier to read and understand to get the timing right than the present digital system of crochets, quavers, semi-quavers etc. A tick up at the beginning of the thick line would indicate a sharp (half note higher) and a tick down a half note lower (a flat). A small space or a mark showing the beginning or the end could separate repeated notes. So far I haven't been able to interest anyone in the system. I have printed an example (a few bars of 'Greensleeves') and it looks fine to me. Only someone in the music publishing business could handle the copyright problems and the promotion expenses. A government could introduce the system in schools but I doubt that any would be motivated. So far I haven't even tried. If a keyboard were to be connected to a printer the notes could match the timing and style of a professional. If the idea were to be taught in schools every child would learn to sing tunes from written music in a few minutes. Playing an instrument would be much easier to learn for beginners. For the idea to be a success, popular music has to be available to read.

Reflecting upon the present state of the world I am inclined to think that now, or soon, might be an appropriate time for God's Day of Judgment! That would solve all our problems.